Understanding
MATHEMATICS

K B Sinha, R L Karandikar, C Musili
S Pattanayak, D Singh and A Dey

Universities Press

UNDERSTANDING MATHEMATICS

UNIVERSITIES PRESS (INDIA) PRIVATE LIMITED

Registered Office
3-6-747/1/A & 3-6-754/1, Himayatnagar, Hyderabad 500 029, Telangana, India
info@universitiespress.com; www.universitiespress.com

Distributed by
Orient Blackswan Private Limited

Registered Office
3-6-752 Himayatnagar, Hyderabad 500 029, Telangana, India

Other Offices
Bengaluru, Chennai, Guwahati, Hyderabad, Kolkata,
Mumbai, New Delhi, Noida, Patna

First published 2000
Reprinted 2003, 2007, 2009, 2024

ISBN 978-81-7371-355-2

501980

Typeset by
Orient BlackSwan

Printed in India at
Graphica Printers & Binders, Hyderabad 500 013

Published by
Universities Press (India) Private Limited
3-6-747/1/A & 3-6-754/1, Himayatnagar, Hyderabad 500 029, Telangana, India

"Mathematics possesses not only truth, but supreme beauty — a beauty cold and austere, like that of sculpture, without appeal to any part of our weaker nature . . . sublimely pure, and capable of a stern perfection such as only the greatest art can show".

Bertrand Russell

Understanding Mathematics

INSA Subcommittee (Mathematics) for Science Education Programme:

Kalyan B. Sinha, F. N. A., F. A. Sc. (Convener)
Indian Statistical Institute, New Delhi, and
Jawaharlal Nehru Centre for Advanced Scientific Research, Bangalore

Rajeeva L. Karandikar, F. A. Sc.
Indian Statistical Institute, New Delhi

C. Musili
University of Hyderabad

Swadhin Pattanayak
Sambalpur University, Orissa

Dinesh Singh
University of Delhi

Aloke Dey
Indian Statistical Institute, New Delhi

Contents

Prologue

The story began with a meeting of the Science Education programme of the Indian National Science Academy on October 11, 1994, chaired by Prof. C.N.R.Rao, in which it was decided to have books written on various science disciplines, not as textbooks but rather as supplementary reading materials. The undersigned as the convener of the writing team on mathematics for senior high school students selected a panel of five other mathematicians for the job. Subsequently, the team met a few times to identify the areas to be covered, the style and points of emphasis in presentation, etc. Though four chapters were ready by the end of 1995, there were some unavoidable delays in completing the other chapters, and work on the full book could start only by the middle of 1997. It was clear to the team right from the beginning that this book will: (i) aim at a readership of the better students at the plus 2 level of the Indian school system, (ii) attempt to explain why the definitions and the theorems are the way they appear, thereby removing some of the mystery behind them, and (iii) have as many examples as possible, drawn mostly from experience. On the other hand, it was recognized that despite the authors' best attempts it may not be possible to maintain uniformity in the style of presentation and in the choice of materials. We also feel that the book will be useful for school teachers and for first year college students.

The team has worked together, read and re-read the whole manuscript, held extensive meetings with and received feedback from a cross section of the mathematical community. We hope the book will be found exciting by school and college students, and that at least some of them will be motivated to take up mathematics as a career.

Kalyan B. Sinha
Indian Statistical Institute,
Delhi Centre,
New Delhi 110 016.

Preface

It is not easy to write a book at a relatively elementary level and yet aim for some amount of completeness. Nevertheless, a serious attempt has been made here to maintain a reasonable degree of completeness. As has been mentioned in the prologue, we have tried to explain why a concept or a theorem came to be the way it is. Of course, this has added to the bulk of the book as well as curtailed sometimes the number of examples. But this is a price that we chose to pay. The choice of topics covered clearly shows a bias towards Analysis—this happened because the school syllabus is so in most cases. The materials are organized in, what we think, a natural order—sets are introduced essentially to bring in the twin concepts of relations and functions, the two most important concepts in modern mathematics. In the second chapter, a brief account of the construction of real numbers is given so that the inquisitive reader gets a flavour of the questions involved. This part may be a bit demanding for a high school student. Chapters 3, 4 and 5 contain mostly standard material, though treated differently. As for example, in Chapter 5, 2×2 matrices have been introduced as linear transformations on 2-dimensional Euclidean space. The sixth chapter on Probability and Statistics is quite non-standard in the sense that it avoids the "mean-mode-median" kind of view of the subject; instead, the reader is guided through situations where asking statistical questions seems the most natural thing to do. The last chapter on algebraic structures and algebraic geometry is the one which is probably the farthest from the standard syllabus. One of the reasons for including this chapter is to dispel a common feeling at this level that mathematics is mostly calculus, the others are to introduce very naturally the concept of a group arising out of the symmetries of geometric figures as well as to give a brief glimpse of the beautiful relationship between geometry and number theory.

Some readers may find a few of the topics difficult at first reading. They are encouraged to skip the proofs, wherever necessary, try to understand the statements of the theorems and proceed to the next topic.

If a few of the readers feel excited about mathematics after reading the book and choose a career in it, then the authors would feel that their efforts have been worth it.

On behalf of the authors, I would like to express my appreciation to INSA for the brave concept of its Science Education Programme and the co-ordinator Dr. Jawahar Dhar for his support. I also thank INSA for providing generous financial assistance. The authors alone are responsible for the views expressed in the book, and for any errors that may have persisted despite their best efforts.

The authors would like to thank the Indian Statistical Institute (Delhi Centre) where most of the organization of the manuscript was done, and the TEX-typists: Mrs. Darshana Bhatia, Mrs. Simmi Marwah and Mr. V. P. Sharma of the Indian Statistical Institute (Delhi Centre) as well as the graphic-artist Mr. Pradip Nayak for their wonderful support and help. We would also like to express our gratitude to our young colleagues Drs. Arup Pal and Anish Sarkar for their help in organizing the final version of the book.

January, 2000

Kalyan B. Sinha
Indian Statistical Institute,
Delhi Centre,
New Delhi 110 016.

Chapter 1

Sets and Functions

"Mathematical language is difficult but imperishable. I do not believe that any Greek scholar of today can understand the idiomatic undertones of Plato's dialogues or the jokes of Aristophanes, as thoroughly as mathematicians can understand every shade of meaning in Archimedes' works."

M.H.A. Newman

1.1 Introduction

Mathematics has been defined as the science of numbers, magnitudes and forms. But, over the years, its scope has so vastly expanded that this definition is hardly comprehensive enough to describe mathematics today. The expansion of its scope has been so much that often two branches of mathematics do not seem to have anything in common to be called the branches of the same discipline. This has grown to the absurd limit of resembling the assertions of the fabled seven blind men trying to describe what an elephant was really like. Can mathematics be presented more *organically*? It *has* been demonstrated that such a scheme of mathematics being put on a common basis is possible. The earliest attempt in this direction was by Georg Cantor at the turn of the 20th century. He tried to show that all concepts of mathematics can be based on a simple idea of "sets". Though there were many flaws in his constructions, the logicians and mathematicians of early twentieth century have been able to sort out the logical difficulties of his methods, and salvage the central theme of his ideas which were about to be consigned to the oblivion of history.

This is an attempt to present mathematics as a contiguous whole based on the improved notion of *sets*. We have all been aware of the notion of sets in our day to day activity. A bunch of keys, a pack of wolves, a swarm of bees, a school of fish, a herd of elephants, a pride of lions, are but a few examples of sets. They are our earliest notions of sets. But a formal definition of sets is really difficult. We *do not* attempt a formal definition of sets. We rather take a set as a primitive notion, so natural, that we need not define it.

In every scheme of logical construction there are some concepts or terms whose meanings (i.e. definitions) we do not explain. They are supposed to be fundamental concepts in terms of which every thing else is described. These are called *undefined* concepts.

Here we take the term *set* to be an undefined term. We think of sets as simple objects. We have given some examples of it already. A *set* is a collection of objects. The objects which constitute the set are called its *members or elements.* For example, an *alphabet* is a set of letters. A cricket team is a set of players. A bunch of keys is a set of keys. A set can be described in two ways. One of the ways is by enlisting its members. If we want to express a set as a list of its elements we usually write all the elements, separated by commas, enclosed within braces as follows:

$$\{a, b, c, d, e, f, g, h, i, j, k, l, m, n, o, p, q, r, s, t, u, v, w, x, y, z\}$$

is the Roman alphabet. Similarly the set $\{1, 2, 3, 4, 5\}$ is a set consisting of 1, 2, 3, 4 and 5 as its elements. The set

is a set of election symbols with

as its members and the set

is a set of figures. Observe that there is no restriction as to which objects can be put together in a list to form a set. What is *most important* in describing a set is that we should be able to say which are members of the set and which are not. For example,

is not a member of the set

whereas

is a member. But the method of enclosing a list of elements in braces is not always possible (or when possible, may not be practical). If we want to express the set of names of citizens of India as a list the task would be stupendous. Even when we are able to write names of 500 Indian nationals in one page of a standard crown size book we would need 1,800,000 pages, and if we have to bind them into books of 500 pages each, then we would have 3,600 books! The reader can easily construct similar examples of sets describing which by a list is impractical. To circumvent this difficulty we have another way of describing a set. In this method we describe a criterion to decide if an object is or is not an element of a set in question. Thus we may write the set of all Indians simply by $\{x : x \text{ is a citizen of India}\}$. Similarly we may denote the set of *integers* (a precise definition of integers can be found in the next chapter) as $\mathbb{Z} = \{n : n \text{ is an integer}\}$. However, we may also write $\mathbb{Z}$ as

$$\mathbb{Z} = \{\cdots, -3, -2, -1, 0, 1, 2, 3, \cdots\}.$$

Observe that in the second way of denoting the set $\mathbb{Z}$ one has to be content with dots in the begining and at the end, as we can never exhaust the list either way, however much we try. In the case of Indian citizens we could at least in principle, write the complete list, however impractical it may be. But in the case of the set of integers, writing the exhaustive list of its elements is *not* possible even in principle, as we can always write a new integer after we have written a list of them. This is the concept of *infinity* to which we shall return later. But when we write an infinite set as a list we have to use dots after writing some elements, when it is clear what its members will be. For example, if we write the set of even numbers as

$$\{\cdots, -6, -4, -2, 0, 2, 4, 6, \cdots\},$$

we know which elements are members of the set though we *cannot* write them all.

Writing a set by a list of all its elements is called defining (or describing) the set by *extension*, whereas writing a set by a criterion or rule which decides the membership of an object is called defining a set by *intention.* The main point is that given a set and an object one can decide if the object belongs to the set or not. Usually, if $p(x)$ stands for the statement about x satisfying some defining property, then the set $\{x : p(x)\}$ is said to be defined by the defining property p. Both the methods of writing a set, by a defining property or as a list are comprehensive enough to include all kinds of sets. A set can be like

{ an ink bottle, a watch, a piece of paper, an eraser, a pencil, a book },

though there is nothing "common" among the members of the set. Suppose my study table has an ink bottle, a watch, a piece of paper, an eraser, a pencil and a book on it and nothing else. I could always describe the above set as the set of things on my study table. The so-called commonality is the fact that they are, at the instant, on my study table. Being on my study table could be the describing property. Alternatively, I can write the set by intention as

$\{x : x =$ an ink bottle or $x =$ a watch or $x =$ a piece of paper,

or $x =$ an eraser or $x =$ a pencil or $x =$ a book$\}$.

Another point we clarify here is that when we are describing a set by a list we do not bother about the order or arrangement in which they are written. Indeed $\{1, 2, 3\}$ stands for the same set as $\{2, 3, 1\}$. This certainly does not affect defining a set by intention. For example,

$\{x : x$ is an animal in the Delhi zoo$\}$

is a set defined by intention but there is no mention of its order.

In dealing with sets we come across a very important set, called the empty set. It has indeed no members as the name suggests. We denote it by $\emptyset$.

The empty set is a very important concept but sometimes confusing. The empty set is a set without any element whatsoever. Then one may wonder why we call it a set at all. We shall see later that we need it in the context of logical completetness. In logic we often come across an idea of something being *vacuously* true. For example, if I promise that I am prepared to donate all my diamond rings to a charitable cause and I have no diamond rings at all then my promise is devoid of any substance in the sense that whatever I do, I won't be accused of having broken my promise. Here if I say D is the set of all the diamond rings I possess, it is clear that D is an empty set. What I am promising is that I would donate every element of D. If D had some elements and I did not donate them then it could be legitimately said that I have broken my promise. But if D has nonc then I could never be accused of breaking my promise.

1.2 Properties of sets

In the context of our dealing with sets we come across a notion called subsets.

DEFINITION: A set A is called a *subset* of a set B if every element of A is an element of B, i.e., if $x \in A$ then $x \in B$. We write this as $x \in A \Rightarrow x \in B$ and express the relation between A and B by

$$A \subset B.$$

Examples

(a) Let $\mathbb{Z}$ be the set of integers. The set of natural numbers $\mathbb{N}$ is a subset of $\mathbb{Z}$.

(b) Let A stand for the set of all animals and M for the set of mammals. It is clear that M is a subset of A.

(c) Let I stand for all Indian nationals and O for the set of all Indian nationals who speak Oriya. O is surely a subset of I.

(d) Let A be the set of Latin alphabets and V be the set of vowels in A. We have $V \subset A$.

(e) Let C be the set of articles in the Indian constitution and F be the set of all those articles which are about fundamental rights. We have $F \subset C$.

Observe that by our definition: *Every set is a subset of itself.* The *empty set is a subset of every set*. This is because there is no element in $\emptyset$ to violate $x \in \emptyset \Rightarrow x \in A$.

If A is a subset of B and B has at least one element which is not in A, then we say that A is a *proper* subset of B. For example, the set $\{a, b, c\}$ is a proper subset of the set $\{a, b, c, d\}$. The set of natural numbers $\mathbb{N}$ is a proper subset of the set of integers $\mathbb{Z}$.

If $A \subset B$ and $B \subset A$ then $A = B$. The very definition of subsets makes this appear obvious. This is in fact taken to be one of the properties of sets. This property is called the *property of extension. This important property* is often used to prove equality of sets.

We have indeed for all X, Y and Z :

1. $X = Y \implies (x \in X \iff x \in Y)$.

2. $X = X$.

3. If $X = Y$ then $Y = X$.

4. $X = Y$, $Y = Z$ implies $X = Z$.

5. $X \subset Y$, $Y \subset Z$ implies $X \subset Z$.

6. $X = Y$ if and only if $X \subset Y$ and $Y \subset X$.

1.3 Union and Intersection of Sets

Suppose a broadcasting organisation seeks to announce in English and Hindi. So they would look for announcers who can speak English and those who can speak Hindi. Suppose E is the set of candidates for the position of announcers who can speak English and H be the set of candidates who can speak Hindi. What is the set of eligible candidates for the job of announcers for the broadcasting corporation? That will be the set of candidates who can speak English or Hindi. This includes all those candidates who can speak both. This set is called the *union* of the sets E and H. It is denoted by $E \cup H$. So if A and B are sets, then

$$A \cup B = \{x : x \in A \text{ or } x \in B\}.$$

The 'or' in the braces includes the case of $x \in A$ and $x \in B$. It is important to mention here a relevant concept called the concept of *exclusive or.* This means only one of the cases $x \in A$ and $x \in B$ is true. But union is *not* defined in terms of exclusive or.

One sees that:

1. $A \subset B \Leftrightarrow A \cup B = B$.

2. $(A \cup B) \cup C = A \cup (B \cup C)$.

3. $A \cup B = B \cup A$.

In this example, if we take the set of candidates who can speak both English and Hindi, then it will be called the *intersection* of the sets E and H. It is denoted by $E \cap H$.

Similarly, suppose an examination has two subjects only, namely mathematics and physics. A candidate has to pass in both the subjects to pass the examination. Now if $M = \{x : x \text{ has passed in mathematics}\}$ and $P = \{x : x \text{ has passed in physics}\}$, then $P \cap M$ is the set of candidates who have passed the examination.

So for sets A and B

$$A \cap B = \{x : x \in A \text{ and } x \in B\}.$$

We have

1. $A \cap B = B \cap A$ for any two sets A and B.
2. $A \cap (B \cap C) = (A \cap B) \cap C$ for any sets A, B and C.
3. $A \cap B \subset A, A \cap B \subset B$ for any sets A and B.
4. If $A \subset B$, then $A \cap B = A$.

If for sets A and B we have $A \cap B = \emptyset$, we say that A and B are *disjoint* or *non-overlapping* sets. That is, two sets are disjoint or non-overlapping if they have no element in common.

1.4 Further properties of sets

For sets A and B we define

$$A \backslash B = \{x : x \in A \text{ and } x \notin B\}.$$

$A \backslash B$ is the set of all those elements of A which are not in B. If E stands for the set of those persons who can speak English and H for the set of persons who can speak Hindi, then $E \backslash H$ would stand for the set of all those persons who can speak English but cannot speak Hindi. Similarly $H \backslash E$ would stand for the set of all those persons who can speak Hindi but not English. The sets $A \backslash B$ and $B \backslash A$ are called the sets of differences. In this context, there is an interesting concept called the *symmetric difference* of two sets. For sets A and B, the set $(A \backslash B) \cup (B \backslash A)$ is called the symmetric difference of A and B and is denoted by

$$(A \backslash B) \cup (B \backslash A) = A \Delta B.$$

For the sets E and H described above, $E \Delta H$ stands for the set of all those persons who can either speak English or speak Hindi but not both. This is the case of exclusive or.

Exercise: In an examination of three subjects physics, chemistry and mathematics, let P stand for the set of those candidates who passed in physics, C for the set of those candidates who passed in chemistry and M for the set of those who passed in mathematics.

Express the following sets by using the symbols of union, intersection and difference.

(a) The set of all those candidates who passed in mathematics only.

(b) The set of all those who passed in mathematics and physics but not in chemistry.

(c) The set of those candidates who failed in one subject only.

(d) The set of those candidates who passed in either physics or chemistry but not in mathematics (we mean or not exclusive or here).

(e) The set of those candidates who passed in all the subjects.

(f) The set of all those candidates who passed at least in two subjects.

In this exercise, if we want to represent the set of those candidates who failed at least in one subject, what do we do? An obvious problem in expressing this as a set is that we have not specified the set of all the candidates who took the examination. Similarly, if we are talking of the set of non-Bengalis we must be clear in which context we are talking? Are we talking of all the people of the world or only about Indian citizens? But when we say "not all birds can fly" we are talking of birds only. We are not talking of the other animals who cannot fly nor are we talking of insects which can fly. Similarly, suppose somebody said "none other than Gandhi could do this feat" we are obviously talking in terms of all human beings. We can find many such examples where it is important to mention the context in which we are talking. In other words we specify a set in terms of subsets of a set about which we are talking. This specified set is sometimes called the "universe of discourse" or the *universal set*. Though we are using the article "the" we are talking of "the universal set" in a particular context.

If X is our universal set and $A \subset X$, then the set $X \backslash A = \{x : x \in X, x \notin A\}$ is called the *complement* of A. It is denoted by A^c. (Strictly speaking we should call $X \backslash A$ as the complement of A with respect to X.) In case X is our universe of discourse we can legitimately claim $X^c = \emptyset$.

DE MORGAN'S LAW One can show that for sets X, A and B we have

(i) $X \backslash (A \cup B) = (X \backslash A) \cap (X \backslash B)$,

(ii) $X \backslash (A \cap B) = (X \backslash A) \cup (X \backslash B)$.

In case X is the universal set of which A and B are subsets we have

(i) $(A \cup B)^c = A^c \cap B^c$,

(ii) $(A \cap B)^c = A^c \cup B^c$.

This is De Morgan's law. The symmetric difference can likewise be written as

$$A\Delta B = (A \cap B^c) \cup (B \cap A^c).$$

De Morgan's law can be illustrated as follows: If A is the set of people who play cricket and B the set of people who play tennis, $A \cup B$ is the set of people who either play cricket or play tennis or play both. Therefore, $(A \cup B)^c$ would mean the set of people who neither play cricket nor play tennis, whereas A^c is the set of people who don't play cricket and B^c the set of people who don't play tennis. So $A^c \cap B^c$ is the set of people who neither play cricket nor tennis. Try the following:

Exercises

Show that for sets A, B and C:

(i) $A \cap (B \cup C) = (A \cap B) \cup (A \cap C)$.

(ii) $A \cup (B \cap C) = (A \cup B) \cap (A \cup C)$.

(iii) $A \cap (B\Delta C) = (A \cap B)\Delta(A \cap C)$.

(iv) $A\Delta A = \emptyset$.

(v) $(A\Delta B)\Delta C = A\Delta(B\Delta C)$.

(vi) $A\backslash(A \cap B) = A\backslash B$.

(vii) $A \subseteq B$ implies $B^c \subseteq A^c$.

Venn Diagram: Often it is very useful to represent sets as regions of the plane (shaded according to convenience).

So if A stands for a region of the plane which is shaded and B for another region of the plane with shades, $A \cup B$ would be the region of the plane with shading (Figure 1.1);

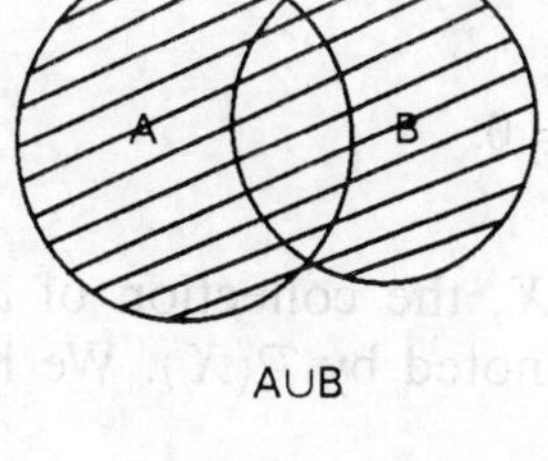

Figure 1.1

whereas $A \cap B$ will be the region of the plane which is common to both A and B (Figure 1.2).

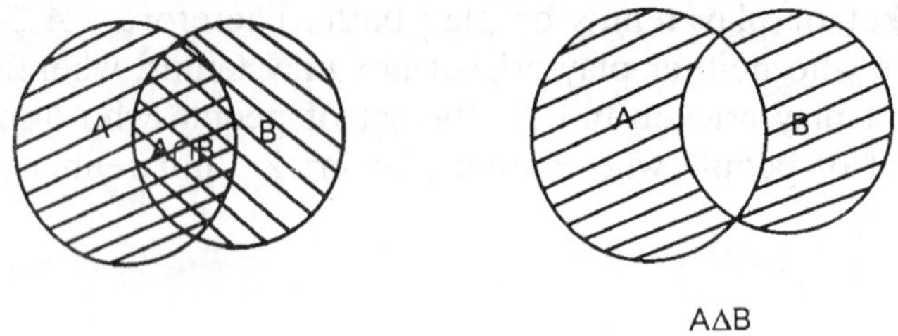

Figure 1.2

Similarly $A\backslash B$ would be the portion of A which is not in B (Figure 1.3).

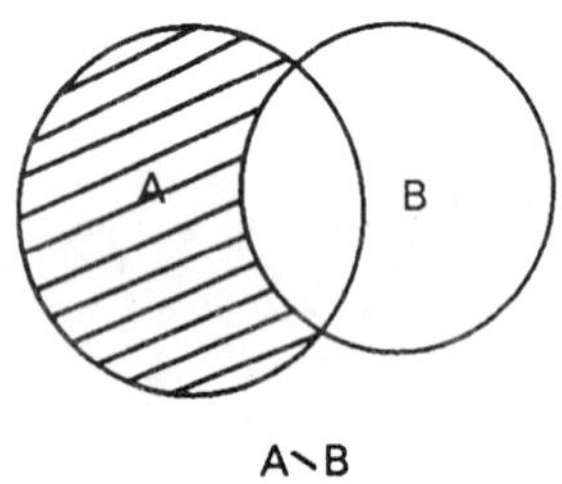

Figure 1.3

Exercises

Show that

(i) $A\backslash B = \emptyset$ if $A \subset B$.

(ii) $A\backslash B = A$ if $A \cap B = \emptyset$.

(iii) $A \Delta B = A \cup B$ if $A \cap B = \emptyset$.

POWER SET: Given a set X, the collection of *all* its subsets is a set which is called the *power set* of X denoted by $\mathcal{P}(X)$. We have

$$\mathcal{P}(X) = \{A : A \subset X\}.$$

Example: If $X = \{a, b, c\}$ we have

$$\mathcal{P}(X) = \{\{a,b,c\}, \{a,b\}, \{b,c\}, \{c,a\}, \{a\}, \{b\}, \{c\}, \emptyset\}.$$

Exercise: Consider the sets

$$\{1,2,3,4\}, \ \{\alpha, \beta, \gamma, \delta, \varepsilon\}, \ \{\oplus, \otimes\},$$

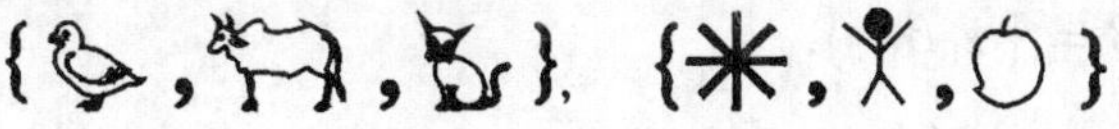

Write the power set of each. Count the number of elements in each set and count the number of elements of the corresponding power set. Do you see any pattern?

1.5 Relation

Among the various kinds of relations we are familiar with, father, mother, son, brother, sister, husband, wife, less than, equal to, parallel to, etc., are a few examples. We would like to fix our ideas as to what exactly is meant by a relation. Let us see how we use the terms in common parlance. We say Dasaratha is the father of Rama. Similarly Babar is the father of Humayun and Janaka is the father of Sita. The above statements like Dasaratha is the father of Rama are but *examples* of fatherhood. But what do we really mean by fatherhood? We may attempt to define fatherhood as the collection of all pairs like Dasaratha and Rama, Babar and Humayun, Janaka and Sita, etc. But in our usage "Dasaratha is the father of Rama". we cannot interchange the *positions* of Dasaratha and Rama. Indeed the statement "Rama is the father of Dasaratha" does not convey the same meaning as the statement "Dasaratha is the father of Rama". Again in the statement "Dasaratha is the father of Rama", if we drop the word Rama, then too it won't mean anything. For the statement about fatherhood, what is important is a *pair* in a definite *order*: (Dasaratha, Rama). Thus a *certain* collection of ordered pairs stands for a definite relation, as in this case the ordered pairs

(Dasaratha, Rama), (Babar, Humayun), (Janaka, Sita)

are examples of relation of fatherhood. So fatherhood would stand for a certain collection of ordered pairs of human beings. Similarly the relation of brotherhood would stand for a certain other collection of ordered pairs. So would sisterhood, etc. In each case they are different collections of ordered pairs. Now we proceed to define the term relation formally. For that we need the notion of cartesian product.

CARTESIAN PRODUCT: For sets A and B, the set of all ordered pairs formed by the first element from A and the second element from B, is called the Cartesian product of A and B. It is denoted by $A \times B$. We have

$$A \times B = \{(a, b) : a \in A \text{ and } b \in B\}.$$

Examples

(a) $A = \{a, b, c, d\}$, $B = \{\alpha, \beta, \gamma\}$.

$$A \times B = \{(a,\alpha),(a,\beta),(a,\gamma),(b,\alpha),(b,\beta),(b,\gamma),(c,\alpha),(c,\beta),$$
$$(c,\gamma),(d,\alpha),(d,\beta),(d,\gamma)\}.$$

(b) $A = \{1, 2, 3\}$, $B = \{$ Ram, Rahim, Peter $\}$.

$A \times B = \{$(1, Ram), (1, Rahim), (1, Peter), (2, Ram), (2, Rahim), (2, Peter),

(3, Ram), (3, Rahim), (3, Peter)$\}$

$A = \{$John, Peter, Edward$\}$, $B = \{$Mary, Jane, Elizabeth$\}$.

$A \times B = \{$(John,Mary), (John,Jane), (John,Elizabeth), (Peter,Mary), (Peter,Jane),

(Peter,Elizabeth), (Edward,Mary), (Edward,Jane), (Edward,Elizabeth)$\}$.

Exercises

(a) Suppose A has n elements and B has m elements, how many elements does $A \times B$ have?
(b) $A = \{\star, 1, a, \emptyset\}$, $B = \{\alpha, \beta, \gamma, \delta\}$. Write $A \times B$.
(c) $A = \{1, 2, 3, 4, 5, 6, 7, 8, 9, 10\}$, $B = \{\{1, 2\}, 3, 4, 5, 6\}$.
What is $A \times B$?
(d) $A = \{?, \$, \Omega\}$, $B = \{\cdot, 1, \oplus\}$. Write $A \times B$.

DEFINITION: By a *relation* or *correspondence* between the elements of a set A and those of a set B, we mean a subset f of the Cartesian product $A \times B$. Thus f is a relation between set A and set B if

$$f \subset A \times B.$$

If $(a, b) \in f$, then we say that *a is related to b through f*. We write this as afb. Thus if H is the set of all human beings and M is the set of all male human beings, then fatherhood is a subset of $M \times H$, whereas brotherhood is another subset of $M \times H$. Motherhood is a subset of $W \times H$, where $W = \{w : w \text{ is a woman }\}$. So if $f \subset M \times H$ is the fatherhood relation, $(a, b) \in f$ means a is the father of b. Similarly if $g \subset M \times H$ is the brotherhood relation, then $(a, b) \in g$ means a is a brother of b. (Observe that a is a brother of b does not imply b is a brother of a since b could be a sister of a.)

DEFINITION: For sets A and B if $f \subset A \times B$, then $\{x : (x, u) \in f \text{ for some } u \in B\}$ is called the *domain* of the relation f, and $\{y : (x, y) \in f \text{ for some } x \in A\}$ is called the *range* of the relation f.

It is quite obvious that the domain of f, above, is a subset of A and the range, a subset of B.

In the above example of fatherhood, its domain cannot contain a woman. Nor need it contain all men. Indeed many men are *not* fathers.

Like father, mother, brother, sister we also have the relation less than, equal to, greater than, similar to, congruent to, perpendicular to, etc.

For the set $A = \{1, 2, 3, 4\}$, let $f \subset A \times A$ be defined by

$$f = \{(1, 2), (1, 3), (1, 4), (2, 3), (2, 4), (3, 4)\}.$$

We see that the relation f here is the relation less than in the set A. Here the domain of f is $\{1, 2, 3\}$ whereas the range is $\{2, 3, 4\}$.

In case $f \subset A \times A$, we call f a relation *in* A. For A the set of all lines 'parallel to' is a subset of $A \times A$. Similarly for the set T of all triangles, similarity and congruence are subsets of $T \times T$. We enumerate a few relations:
(Remember, afb means $(a, b) \in f$.)

1. A relation f in A (i.e., $f \subset A \times A$) is called a *symmetric* relation if

   $afb \Longleftrightarrow bfa.$

2. $f \subset A \times A$ is called a *reflexive* relation if

   afa for all $a \in A$.

3. $f \subset A \times A$ is called an *antisymmetric* relation if

   afb and $bfa \Rightarrow a = b.$

4. A relation $f \subset A \times A$ is called a *transitive* relation if

   $afb,\ bfc \Rightarrow afc.$

Examples

(a) Similarity of triangles is reflexive, symmetric and transitive.

(b) Fatherhood is neither reflexive, nor symmetric nor transitive.

(c) "Ancestor of" is transitive but is neither reflexive nor symmetric.

(d) The relationship "a is the spouse of b" is symmetric but neither reflexive nor transitive.

(e) "Less than" is transitive but not reflexive.

(f) "Less than or equal to" is reflexive, antisymmetric and transitive.

(g) If X is a class of sets, let $R \subset X \times X$ be defined by ARB if $A \subset B$. We see that R is a reflexive, antisymmetric and transitive relation.

Digression: The way we have defined relation should strictly be called *binary* relation as it consists of ordered *pairs*. But we have other kinds of relations too. Like collinearity of points, concurrence of lines, etc. They are *not* binary relations. But we don't go into them here.

Diagramatic Representation of Relations

A relation between elements of a set A and those of a set B can be represented by diagrams as shown below.

The elements of the sets A and B can be represented by points (in our case in a plane). If $(a, b) \in f$, then we draw a line with an arrow connecting a to b pointing towards b. The collection of all such lines with arrow heads shall be the relation f.

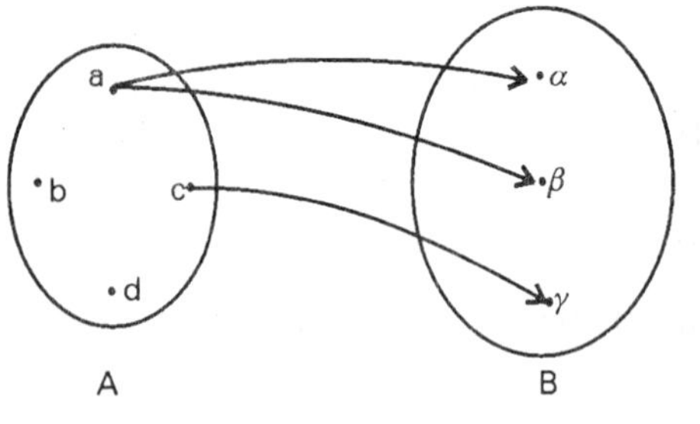

Figure 1.4

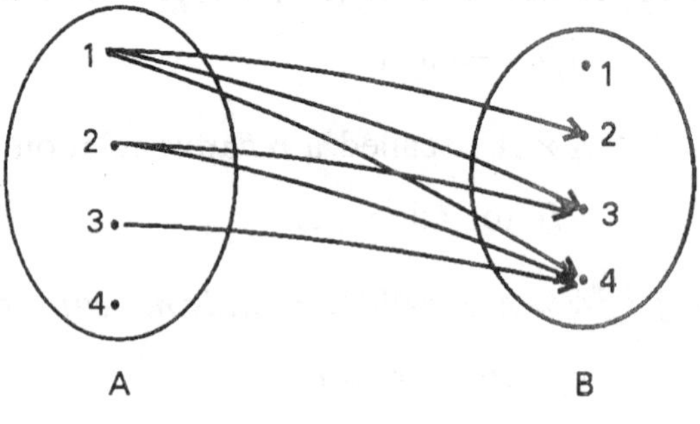

Figure 1.5

In Figure 1.4, $A = \{a, b, c, d\}$, $B = \{\alpha, \beta, \gamma\}$ and the picture represents the relation $\{(a, \alpha), (a, \beta), (c, \gamma)\}$.

Figure 1.5 represents the relation $\{(1,2),(1,3),(1,4),(2,3),(2,4),(3,4)\}$.

If for a relation $f \subset A \times B$, $(a,b) \in f$ and $(a,b') \in f$ and $b \neq b'$, then we call f a *one-many* relation. In the case of one-many relations there are lines from a single element of A to more than one element in B. Similarly, for $f \subset A \times B$ if $(a,b) \in f, (a',b) \in f$ and $a \neq a'$ we call f a *many-one* relation. In such a case there will be lines starting from more than one element in A and ending at the same element in B.

If for $f \subset A \times B$, (i) (a,b) and $(a,b') \in f \Rightarrow b = b'$; and (ii) $(a',b),(a,b) \in f \Rightarrow a = a'$; then we call f a *one-to-one* relation. We may say that a one-to-one relation is neither one-to-many nor many-to-one (see Figure 1.6)

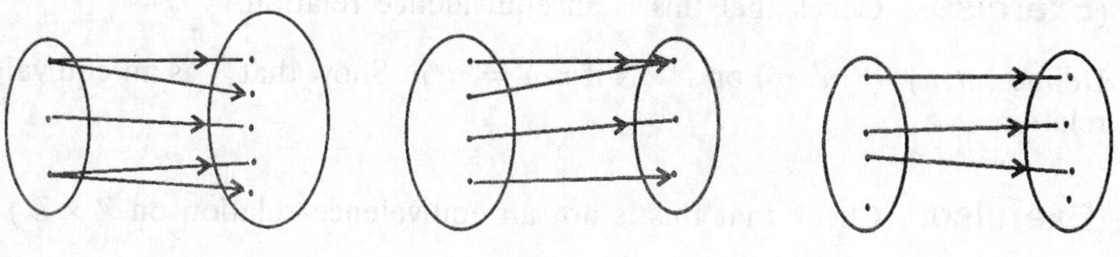

one-many many-one one-to-one

Figure 1.6

For example, fatherhood clearly is a one-many relation whereas the relation of being a son is a many-one relation. Brotherhood is both one-many and many-one. Husbandhood or wifehood (if there is no polygamy or polyandry) is a one-to-one relation. "Less than" is both one-many and many-one. Equality is clearly one-to-one.

Equivalence relation: A relation $\sim$ on A is called an equivalence relation if it is reflexive, symmetric and transitive, i.e.,

(i) $a \sim a$ for all $a \in A$.

(ii) $a \sim b \Leftrightarrow b \sim a$.

(iii) $a \sim b$ and $b \sim c \Rightarrow a \sim c$.

Examples

(a) Equality is certainly an equivalence relation.

(b) Similarity between triangles is an equivalence relation.

(c) Congruence of triangles is an equivalence relation.

(d) Parallelism between lines is an equivalence relation if we agree that a line is parallel to itself.

(e) Define a relation '$\equiv$ (mod 5)' on $\mathbb{Z}$ by $m \equiv n \pmod 5$ if $m - n$ is divisible by 5. Note that this is an equivalence relation.

(f) Let $A = \mathbb{N} \times \mathbb{N}$, where $\mathbb{N}$ stands for the set of all natural numbers. Define $\sim$ on A by

$$(m, n) \sim (m', n') \quad \text{if} \quad m + n' = m' + n.$$

(**Exercise**: Check that this is an equivalence relation.)

(g) Define $(m, n) \sim (m', n')$ on $\mathbb{N} \times \mathbb{N}$ if $mn' = m'n$. Show that $\sim$ is an equivalence relation.

(**Exercise**: Check that this is not an equivalence relation on $\mathbb{Z} \times \mathbb{Z}$.)

Equivalence class: Let X be a set with an equivalence relation $\sim$. For $x \in X$, we write $[x]$ for the set

$$\{y : x \sim y\}$$

which is the set of all the elements of X that are related to x. This is called the *equivalence class* of x.

It is easy to see that if

$$[x] \cap [y] \neq \emptyset, \text{ then } [x] = [y].$$

That is to say that two equivalence classes are either disjoint or are identical. Besides, every element of X is in some (and hence a unique) equivalence class. This can be rephrased as: *An equivalence relation on a set X decomposes X into disjoint equivalence classes.* For example, if we take '$\equiv$ (mod 5)' as the equivalence relation on $\mathbb{Z}$, then there are exactly five equivalence classes:

$$\begin{aligned}
[0] &= \{\cdots, -15, -10, -5, 0, 5, 10, 15, \cdots\} \\
[1] &= \{\cdots, -14, -9, -4, 1, 6, 11, 16, \cdots\} \\
[2] &= \{\cdots, -13, -8, -3, 2, 7, 12, 17, \cdots\} \\
[3] &= \{\cdots, -12, -7, -2, 3, 8, 13, 18, \cdots\} \\
[4] &= \{\cdots, -11, -6, -1, 4, 9, 14, 19, \cdots\}.
\end{aligned}$$

Observe that

$$\begin{array}{lllll}[0] &=& [5] &=& [10]\cdots,\text{similarly}\\ [-4] &=& [1] &=& [6]\cdots,\\ [-3] &=& [2] &=& [7]\cdots,\\ [-2] &=& [3] &=& [8]\cdots,\\ [-1] &=& [4] &=& [9]\cdots.\end{array}$$

Exercises

(i) Given any integer $k > 0$, define $a \equiv b \pmod{k}$ by $a \equiv b \pmod{k}$ if $b - a$ is divisible by k. Show that $\equiv \pmod{k}$ is an equivalence relation. How many distinct equivalence classes does this decompose $\mathbb{Z}$ into?

(ii) Let Σ consist of the points of the plane minus the point 0. For $p_1, p_2 \in \Sigma$, define $p_1 \sim p_2$ if we can obtain p_2 by rotating p_1 about 0 through an angle. Show that $\sim$ is an equivalence relation in Σ. What are the equivalence classes of this?

For a set X and $\sim$ an equivalence relation on X, we define the set of equivalence classes

$$\{[x] : x \in X\}$$

as the quotient set X modulo the equivalence relation $\sim$. It is denoted by $X/\sim$. That is if $[x] = \{y : y \sim x\}$, then

$$\{[x] : x \in X\} = X/\sim.$$

Order relation: We call a relation R in a set X a *partial ordering* if

(i) aRa for every $a \in X$,

(ii) $aRb, bRa \Rightarrow a = b$,

(iii) $aRb, bRc \Rightarrow aRc$.

That is, a relation which is reflexive, anti-symmetric and transitive is called a *partial ordering*. A set together with a partial ordering is called a *partially ordered* set or a *poset*. It is customary to denote a partial order by $\leq$.

Examples

(a) If X is a set of sets, the inclusion relation $\subset$ is a partial ordering.

(b) On the set $\mathbb{Z}_+$ of positive integers define divisibility as follows: for positive integers a and b, $a|b$ if b is divisible by a. One can easily show that $|$ is a partial ordering in $\mathbb{Z}_+$.

(c) Let $\mathbb{Z}$ be the set of integers. Let $\leq$ be defined by

$a \leq b$ if either $a < b$ or $a = b$. Then $\leq$ is a partial ordering in $\mathbb{Z}$.

DEFINITION: Suppose X is a set with a partial ordering $\leq$. We say two elements $a, b \in X$ are *comparable* if either $a \leq b$ or $b \leq a$. If $\leq$ is such that any two elements of X are comparable, then we say that $\leq$ is a *linear ordering* or a *total ordering*.

Observe that in the example above, the inclusion relation is *not* a linear ordering. Similarly, divisibility is a partial ordering but not a total ordering, whereas less than or equal to in $\mathbb{Z}$ is a linear ordering.

Exercise: Consider $V = \{w : w$ is a word in the English language $\}$. We compare the words w_1 and w_2 as follows: $w_1 \leq w_2$ if the first letter of w_1 occurs before the first letter of w_2 in the roman alphabet. If the first letter of w_1 is the same as the first letter of w_2, then consider the second letters of w_1 and w_2. If the second letter of w_1 occurs before that of w_2 in the roman alphabet, then $w_1 \leq w_2$. If the second letters too are the same, we go to the third letters and compare them accordingly. If, however, the number of letters in w_1 is less than the number of letters of w_2, w_1 has n letters and the first, second..., nth letters of w_2 are those of the letters of w_1 respectively, then $w_1 \leq w_2$. Show that $\leq$ orders V linearly. (Such an ordering is called dictionary or lexicographic ordering.)

Exercise: Let $I_2, I_3, I_4, \ldots, I_m$ ... be the subsets of $\mathbb{Z}$ defined by

$$\begin{aligned}
I_2 &= \{2n : n \in \mathbb{Z}\} &= \{\cdots, -4, -2, 0, 2, 4, \cdots\} \\
I_3 &= \{3n : n \in \mathbb{Z}\} &= \{\cdots, -6, -3, 0, 3, 6, \cdots\} \\
I_4 &= \{4n : n \in \mathbb{Z}\} &= \{\cdots, -8, -4, 0, 4, 8, \cdots\} \\
&\vdots \\
I_m &= \{mn : n \in \mathbb{Z}\}.
\end{aligned}$$

The class $\mathcal{I} = \{I_2, I_3, I_4, \ldots\}$ is surely partially ordered by inclusion.

Exercise: Identify those elements of $\mathcal{I}$ that are not subsets of any other element of $\mathcal{I}$.

1.6 Function

Among the kinds of relations we have been talking about, there are some which are very useful in mathematics. A relation $f \subset A \times B$ is called a *function* from A to B if

(i) domain of $f = A$,
(ii) $(a, b) \in f$ and $(a, b') \in f \Rightarrow b = b'$.

Thus a function from A to B is a relation whose domain is the whole of A and which is not one-to-many. The set B is called the *codomain* or *range* of the function f.

If $f \subset A \times B$ is a function from A to B, then we write $f : A \to B$ and say that f is a function from A to B. We also say that f *maps* A into B and often a function is called a *map* or *mapping*. If $(a, b) \in f$, we write

$$b = f(a).$$

We call b the *value* of the function f at a or the *image* of a under f. Observe that there can be no ambiguity in writing $f(a)$ because it is impossible that $(a, b) \in f$ and $(a, b') \in f$ and $b \neq b'$. So by definition of a function $f : A \to B$, we have for every $a \in A$ a *unique* $f(a) \in B$. Thus a function would be completely determined if we knew $f(a) \in B$ for every $a \in A$. That is why sometimes a function $f : A \to B$ is defined as a "rule"which associates to every element $a \in A$ a unique element $f(a)$ of B. We are obliged to put the word *rule* within quotation marks for the simple reason that the rule is often elusive. In fact, the object in many situations is to "discover" this rule.

Examples

(a) Suppose we are observing the position of a particle moving in a straight line. We know that at every instant of time the particle has a unique position on the line. If we agree that every point of the line corresponds to a real number with some convenient point on it decided to represent the real number 0, then the position of the particle at the time t is represented by the real number $x(t)$. We observe that with our ordinary notion of time and position, we do not expect a particle to occupy two different positions at the same time (nothing prevents, though, the particle from occupying the same position at two different times. Indeed, if the particle is at rest for a certain length of time then it would have the same position for the entire length of time). Suppose every time is represented by a real number, 0 being a certain time deemed as the present time, then all future times are represented by positive real numbers and past times by negative real numbers. Here, a real number is used as an intutive object, say, as points on a line, postponing a mathematical discussion of real numbers till Chapter 2. Let $\mathbb{R}$ be the set of real numbers; then the position of a particle moving in a straight line is, in fact, a function

$$x : \mathbb{R} \to \mathbb{R}$$

which associates to every time t the position $x(t)$.

(b) Look at a railway time table. Below a particular train there are times recorded in a column. It records the times of arrival and departure of a train at certain stations. We may look upon this as a function whose domain consists of disjoint intervals of time and range consists of certain cities where the train stops. Thus for every time in the mentioned interval we have a city where the train is at that time. The table does not say anything about the train's position at a time after its departure from a station and before its arrival at the next. This does not disqualify it to be a function if we take the domain to be the intervals of time depicting the times of arrival and departure at certain stations.

(c) Consider the distance of a particle, falling freely under gravity, measured from the point from where the particle started its fall. If we record the time since it began its fall, we know that the distance $x(t)$ at a time t is given by the formula

$$x(t) = \frac{1}{2}gt^2$$

(assuming, of course, that the gravity does not vary and there is no air resistance). Here we *have* in fact a *rule* which tells us where the particle would be at any time. This certainly represents a function. However we are not often so lucky to have a neat formula like this for other functions.

(d) The morning news on a particular day tells us the temperature of the four metros recorded at 5:30 a.m. on that day. Here we have a definite temperature for every such metro at 5:30 a.m. of that day. So it is in fact a function whose domain consists of the set of four metros and range the real numbers quantifying temperature.

(e) Look at your school time table. What does it record on a particular day? It records the subjects to be taught at different times of the day. A particular student has a particular subject being taught during a particular period, as a student is not expected to be in two different classes at the same time. So for each student the time table is a function whose domain is the set of periods and the range the set of subjects.

(f) When we want to mail a letter enclosed in an envelope, we usually go to the post master with it to tell us the denomination of the stamp to be affixed. The post master weighs the letter and tells us the postage according to the weight. Here we have a definite postage for a definite weight, we don't have different rates for the same weight (for same kind of mail like registered or ordinary). The rate chart with the post master, for a particular kind of mail, truly, is a function whose domain is the set of positive real numbers representing the weights of the mail, and the range again consists of postive real numbers representing the postage. We write the chart as a function f such that

$$p = f(w),$$

meaning, p is the postage to mail a letter of weight w.

(g) We know that solubility of a salt varies with temperature. That is to say that a given salt has a definite solubility at a definite temperature. So the relation which associates to every temperature the definite solubility of a salt is in fact a function. Sometimes we say that the solubility is a function of temperature.

(h) Consider the population of a community of a biological species at different times. If $N(t)$ is the population of the community at time t then surely

$$N : \mathbb{R} \to \mathbb{R}$$

represents a function.

(i) If we measure the voltage of an A.C. power supply over a period of time, we observe that it is different at different times. At every instant of time it has a voltage. Sometimes it is positive, sometimes negative and sometimes zero. In fact, for ordinary domestic supply in India the voltage varies between 330 volts and −330 volts. But this change from 330 volts to −330 volts occurs within an interval of 1/100th of a second. (The ordinary D.C. voltmeters of your laboratory won't be able to record this quick change in voltage.) In fact, if $V(t)$ is the voltage at time t, we have

$$V(t) = E_0 \sin(wt + \varphi),$$

where $w = \frac{2\pi}{50}$ and $E_0 = 330$, and φ is a constant known as the phase.

(j) Let S be a sphere resting on a horizontal plane H. Let n be the north pole and s be the southpole, where the sphere is touching the plane (see Figure 1.7). Now let us join any other point $p \in S$ to n and extend it to meet the plane at some point q. Now to every point $p \in S - \{n\}$, we have a unique point q on H. This defines a function $f : S - \{n\} \to H$. This function is called the stereographic projection of the sphere on the plane.

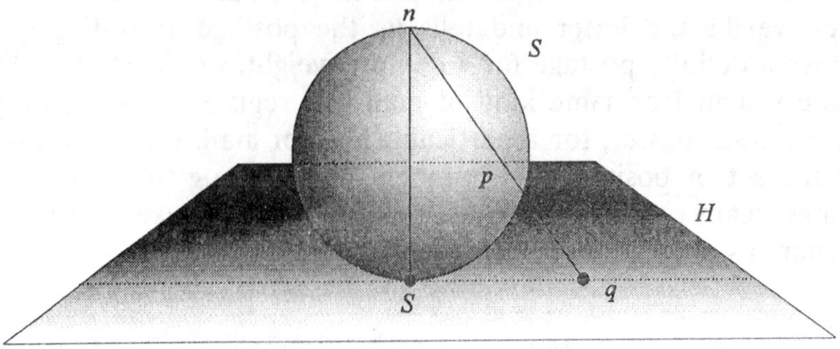

Figure 1.7

(k) The trivial map or function $id : A \to A$ which sends every element of the set A to itself is called the *identity* map.

Exercise: Give 5 more examples of functions you know in real life situations.

For a function $f : A \to B$ and $E \subset A$, we write $f(E) = \{f(a) : a \in E\}$ and call it the image of the set E under f. Also, $f(A)$ is called the range of the function f.

Example: Let $f : \mathbb{R} \to \mathbb{R}$ be defined by $f(x) = \sin x$. It is clear that the image of f is the interval $\{x : -1 \leq x \leq 1\}$. If $E = \{x : 0 < x < \pi\}$, then $f(E) = \{x : 0 < x \leq 1\}$.

1.7 Graph of Functions

In the case of a function whose domain is a subset of the set of real numbers and the range is the set of real numbers, we can represent it by a graph in $\mathbb{R}^2$, the Cartesian product of $\mathbb{R}$ with itself, which is often called the Euclidean plane. Indeed, as the function is a collection of ordered pairs, the collection of $\{(x, y) \in f : x, y \in \mathbb{R}\}$ is a set of points in the plane $\mathbb{R}^2$.

We recall that once we fix the coordinate axes, a point in the plane can be represented by an ordered pair of real numbers and conversely, every ordered pair of real numbers represents a point on the plane. Thus if we have a function with domain and range as subsets of the set of real numbers, then the function corresponds to a set of points in the plane, the domain of the function being a subset of the abscissa and the range a subset of the ordinate. This collection of points $\{(x, y) \in f : x, y \in \mathbb{R}\}$, which is simply the relation defining f, is called the *graph* of the function f. If $f : A \to \mathbb{R}$ is a function and $A \subset \mathbb{R}$ then its graph is

$$\{(x, f(x)) : x \in A\}.$$

Examples

(a) Motion of a particle:

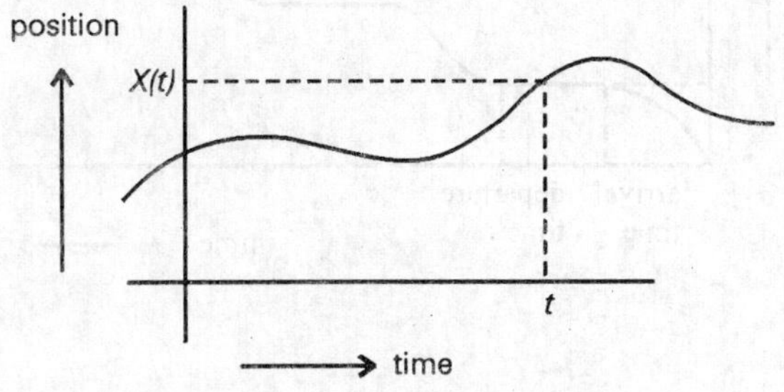

Figure 1.8

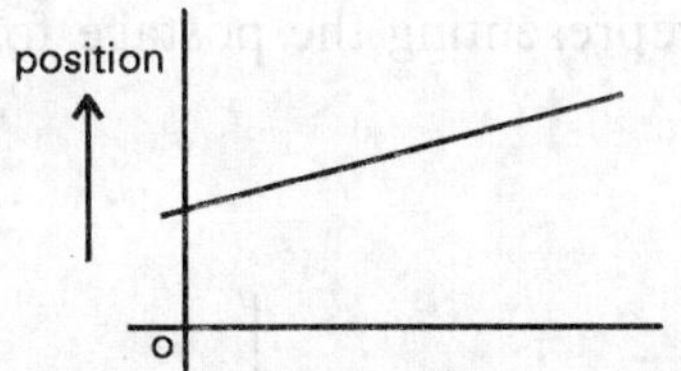

Figure 1.9

The picture in Figure 1.8 shows the position of a particle at a given time t. If it is the case of a particle moving with uniform velocity, then its graph is shown as a straight line (Figure 1.9).

For a particle falling freely under gravity from rest, the distance being measured from the point of its falling, then its graph is given by the picture as in Figure 1.10.

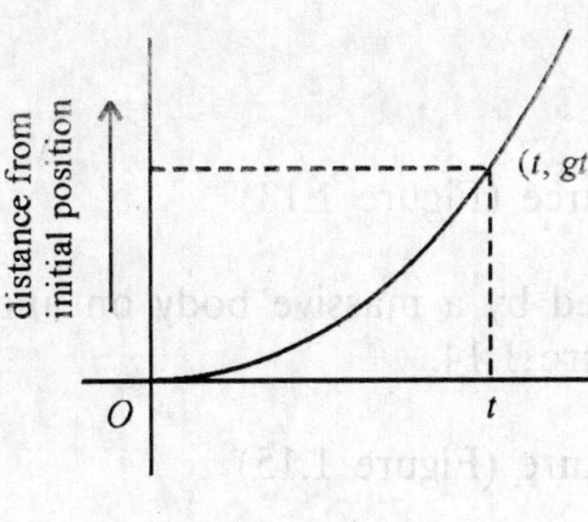

Figure 1.10

In case it is a function representing the distance of a train from the station of its starting, then its graph would be like in Figure 1.11.

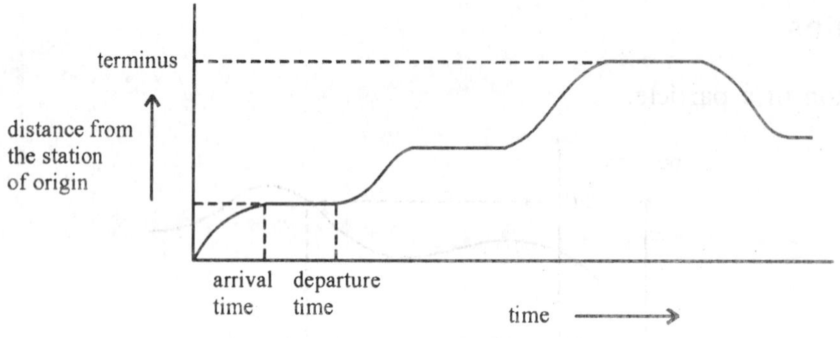

Figure 1.11

(b) Postage versus weight:

The graph of the function representing the postage for different weights will be as in Figure 1.12.

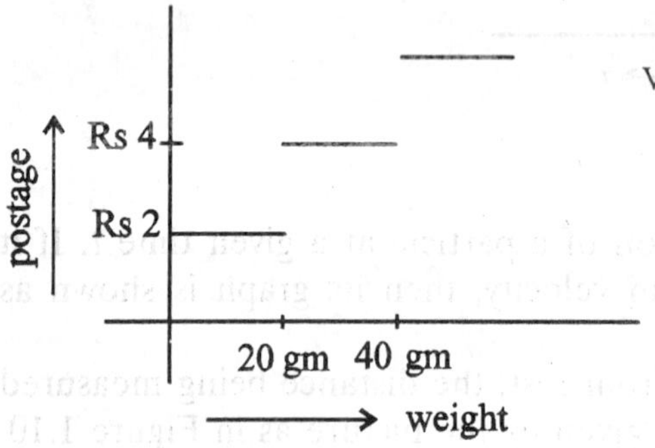

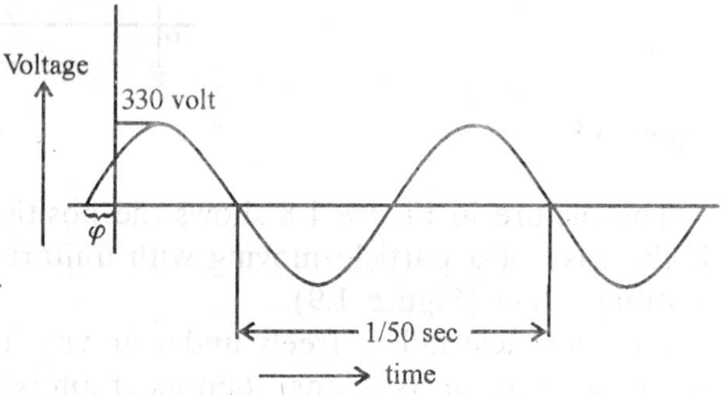

Figure 1.12 **Figure 1.13**

(c) Voltage from an A.C. source (Figure 1.13):

(d) Force of attraction exerted by a massive body on another body at a distance has a graph as shown in Figure 1.14.

(e) Solubility versus temperature (Figure 1.15)

(f) Biological population, with unrestricted growth (Figure 1.16)

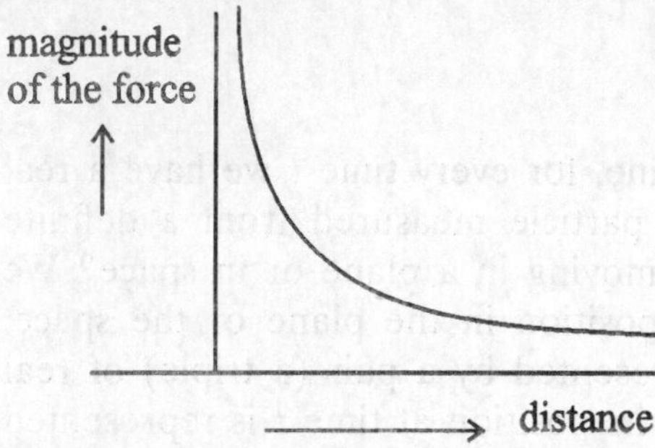

Figure 1.14

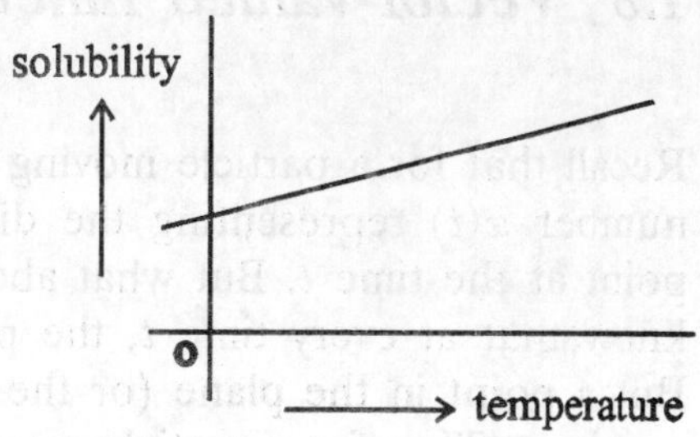

Figure 1.15

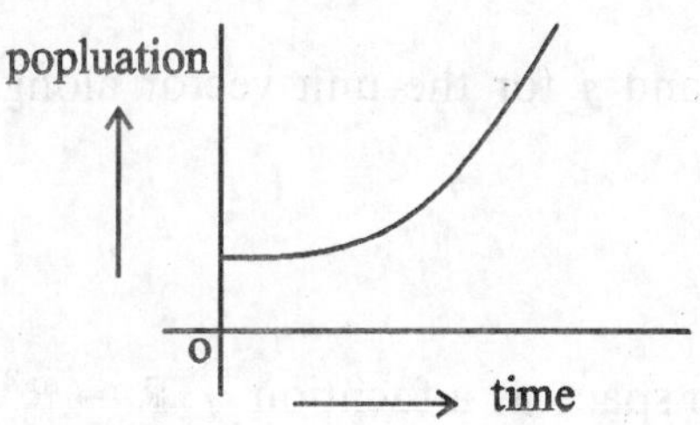

Figure 1.16

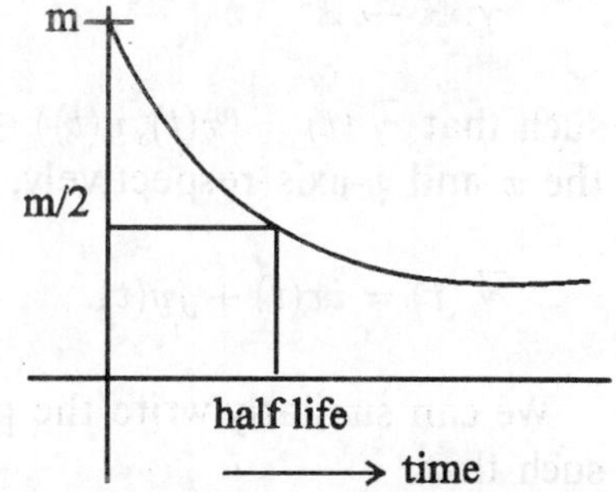

Figure 1.17

(g) Mass of a radioactive substance versus time:

We know that radioactive substances decay over time, i.e. as time passes the mass of a lump of radioactive substance decreases. The graph of the function representing its mass at different times is given by Figure 1.17.

Exercise: Draw graphs of 5 more functions taken from real life situations.

(h) Distribution of prime numbers:

Let $\Pi(x)$ be the number of prime numbers less than or equal to the real number x. Clearly Π is a function from $\mathbb{R}$ to $\mathbb{N}$, and for example, $\Pi(x)$ is zero for $0 \leq x < 2$ (recall that the number 1 is not deemed a prime), equals 1 for $2 \leq x < 3$, etc., equals 4 for $7 \leq x < 11$ and so on. The reader is encouraged to draw the graph of Π for $x \leq 1000$. Till now, there is no formula for Π; however, Hardy and Ramanujan obtained its behaviour for large x.

1.8 Vector-valued functions

Recall that for a particle moving in a straight line, for every time t we have a real number $x(t)$ representing the distance of the particle measured from a definite point at the time t. But what about a particle moving in a plane or in space? We know that at every time t, the particle has a position in the plane or the space. But a point in the plane (or the space) is represented by a pair (a triple) of real numbers. Thus for a particle moving in a plane its position at time t is represented by a pair $(x(t), y(t))$. We may say that the pair is the value of a function whose range is the set of points in the plane $\mathbb{R}^2 (\equiv \mathbb{R} \times \mathbb{R})$. Thus the function representing the position of the particle at different times is a function

$$\vec{\gamma}: \mathbb{R} \to \mathbb{R}^2$$

such that $\vec{\gamma}(t) = (x(t), y(t)) \in \mathbb{R}^2$, or if we write $\hat{\imath}$ and $\hat{\jmath}$ for the unit vector along the x and y-axis respectively, then

$$\vec{\gamma}(t) = \hat{\imath}x(t) + \hat{\jmath}y(t).$$

We can similarly write the position of a particle in space as a fucntion $\vec{\gamma}: \mathbb{R} \to \mathbb{R}^3$ such that

$$\vec{\gamma}(t) = \hat{\imath}x(t) + \hat{\jmath}y(t) + \hat{k}z(t).$$

Function of many variables

We have discussed before that the temperature at a point on Earth, at any instant, is a unique real number. Now every point on Earth is represented by a pair of real numbers depicting its latitude and longitude respectively (one ought to be careful in making this statement when it comes to a point on the date line. Indeed there is a little ambiguity in representing the longitude of a point on the date line. Besides, the poles have unique latitude, but what about their longitudes? Barring such ambiguity, every point on Earth can be represented uniquely by a pair of real numbers). Thus for a point with latitude θ and longitude φ, we have a definite temperature $T(\theta, \varphi)$ at any instant. Thus,

$$T: \left(-\frac{\pi}{2}, \frac{\pi}{2}\right) \times (0, 2\pi) \to \mathbb{R}$$

represents the temperature at a point.

Similarly, if we take any point in the atmosphere, the atmospheric pressure at the point depends on the latitude, the longituide and the altitude of the point.

Indeed, for a point with latitude θ, longitude φ and altitude h, at a given instant, we have a unique real number $p(\theta, \varphi, h)$ called the atmospheric pressure at that point. Thus p can be deemed as a function whose domain is a part of $\mathbb{R}^3$ and range $\mathbb{R}$.

VECTOR FIELDS: Electric field strength at a point is defined as the force experienced by a unit electric charge at that point. This means that with every point $(x, y, z) \in \mathbb{R}^3$ is associated a vector $\vec{E}\,(x, y, z) \in \mathbb{R}^3$ which is called the electric field strength at the point (x, y, z). Thus we may think of electric field strength as a map $\vec{E}\colon \mathbb{R}^3 \to \mathbb{R}^3$. Similarly, magnetic field strength is a function $\vec{H}\colon \mathbb{R}^3 \to \mathbb{R}^3$ and the velocity of a fluid is again a function (or map) $\vec{q}\colon \mathbb{R}^3 \to \mathbb{R}^3$.

Exercises

(a) Give five more examples of vector-valued functions.
(b) Give five more examples of functions of many variables.

1.9 Composition of functions

Suppose A, B and C are sets and $f : A \to B$ and $g : B \to C$ are functions. We define a function

$h : A \to C$ by

$h(a) = g(f(a))$ for every $a \in A$.

It is easily seen that h is a well defined function, as $f(a) \in B$ for every $a \in A$ and $g(b) \in C$ for every $b \in B$. The function h is called the *composition* of the functions g and f and is denoted by $g \circ f$.

Examples

(a) Suppose that in a forest, carnivorous animals sustain themselves by feeding only on herbivorous animals and that the nutrition level of a herbivorous animal depends on the vegetation around the animal. So the nutrition level of a carnivorous animal ultimately depends on the vegetation around the population of herbivorous animals it feeds on. Thus if V is the density of vegetation around the herbivorous animals and n is the level of nutrition of the herbivorous animal (for simplicity measured by its weight, though there are often more parameters depicting the level of nutrition of an animal), $n : V \to \mathbb{R}$ is a function. Similarly if c is the level of nutrition of a carnivorous animal, then c is a function depending on the level of nutrition of herbivorous animals it feeds on:

$$c : \mathbb{R} \to \mathbb{R}.$$

Thus

$$c \circ n : V \to \mathbb{R}$$

is the level of nutrition of the carnivorous animal ultimately depending on the density of vegetation.
(b) Take for example the force experienced by a moving charged particle in a magnetic field which is varying in time. We know that the force on the charged particle depends on the strength of the magnetic field in which it moves. Again, as the magnetic field strength varies with time, the force experienced is ultimately a function of time and position.
(c) Suppose there is a lamp in the room. The intensity of illumination at a point in the room depends on the illuminating power of the lamp. The illuminating power of the lamp again depends on the voltage of the electric power supply which makes the lamp glow. So ultimately the intensity of illumination depends on the voltage of the power supply.

Exercise: Give five more examples of composition of functions.

1.10 Injections, surjections and bijections

A function $f : A \to B$ is called an *injection* if $f(a) = f(a')$ implies $a = a'$. There is another name for this kind of function. It is also called a *one-to-one function* or an *injective map*. A map $f : A \to B$ is one-to-one if two distinct elements of A have *distinct* images under f.

A function $f : A \to B$ is called a *surjection* if for every $b \in B$ there is an element $a \in A$ such that

$$f(a) = b.$$

In other words, $f(A) = B$. That is to say every element of B *is* an image of some element of A under f. A surjective map is also called an *onto* map.

A map $f : A \to B$ which is both one-to-one and onto is called a *bijection* or a *bijective map*.

Exercises

Which of the functions defined below are one-to-one and which are onto? Which are both? Given $f : \mathbb{R} \to \mathbb{R}$ defined by:

(i) $f(x) = x^2$, (ii) $f(x) = x^2 + x + 1$, (iii) $f(x) = \sin x$, , (iv) $f(x) = x^3$,
(v) $f(x) = \tan x$ for $x \neq (2n+1)\frac{\pi}{2}$, $f((2n+1)\frac{\pi}{2}) = 0$, (vi) $f(x) = e^x$,
(vii) $f(x) = \frac{e^x - e^{-x}}{2}$.

Exercises

(i) Show that if for a map $f : A \to B$ there exists a map $g : B \to A$ such that $f \circ g(y) = y$ for every $y \in B$, then f is onto.
(ii) Show that if for $f : A \to B$ we have $g : B \to A$ such that $g \circ f(x) = x$ for all $x \in A$, then f is one-to-one.

Examples

1. Suppose $f : \mathbb{R} \to \mathbb{R}$ is defined by $f(x) = \cos x$. It is clear that this is neither one-to-one nor onto. Indeed because $f(x) = f(x + 2\pi)$, it cannot be one-to-one. Since $f(x)$ never takes a value below -1 or above 1, it cannot be onto.

2. $f : \mathbb{R} \to \mathbb{R}$ defined by
$$f(x) = \begin{cases} x & \text{if } x \geq 0, \\ x - 1 & \text{if } x < 0. \end{cases}$$
is one-to-one but not onto as $-1/2$ is never attained by the function.

3. $f : \mathbb{R} \to \mathbb{R}$ defined by $f(x) = \frac{x}{1+|x|}$ is one-to-one but not onto.

Exercises

For the functions defined below, determine which are one-to-one, which are onto, which are both and which are neither.

(i) $f : (-1, 1) \to \mathbb{R}$, $f(x) = \dfrac{x}{1 - x^2}$.

(ii) $f : \mathbb{R} \to \mathbb{R}, \quad f(x) = x^6.$

(iii) $f : (0, 2\pi) \to [-1, 1], \quad f(x) = \sin x.$

(iv) $f : \left[0, \frac{\pi}{2}\right] \to [0, 1], \quad f(x) = \sin x.$

Exercise: Define three more functions $f : \mathbb{R} \to \mathbb{R}$ which are both one-to-one and onto.

Let $f : A \to B$ be a function. Let us define

$$f^{-1} = \{(b, a) : (a, b) \in f\} \subset B \times A.$$

Surely f^{-1} is a relation. Now for f^{-1} to be a function it is clear that the domain of f^{-1} should be B, which is the same thing as asking that f be surjective, and that $(b, a), (b, a') \in f^{-1}$ should imply $a = a'$. This is the same thing as saying $f(a') = f(a) = b$ should be true only when $a' = a$. So f^{-1} is a function when f is a bijection, f^{-1} is called the *inverse* function of f and it is easy to see that $f^{-1} \circ f = f \circ f^{-1} = id$, the identity map.

Warning: Trigonometric functions like sine and cosine are neither one-to-one nor onto. So how does one define $\sin^{-1} x, \cos^{-1} x$? Actually there is ambiguity in defining these. If we write $\sin^{-1} x = \theta$, it means that $\sin\theta = x$. It is easily seen that there is no θ if x is more than 1 or less than -1. Thus, the domain of $\sin^{-1}$ or $\cos^{-1}$ must be [-1, 1]. Then again, $\sin\theta = x$ has many solutions θ for the same x. For example, $\sin(\pi/6) = \sin(5\pi/6) = \frac{1}{2}$. So which of $\pi/6$ or $5\pi/6$ should claim to be the value of $\sin^{-1}(1/2)$? In such a case, we agree to take only one value in a definite way. For $0 \le x \le 1$ we choose $0 \le \theta \le \pi/2$ such that $\sin\theta = x$. It is obvious that there is only one such θ. Similarly for $-1 \le x < 0$ we choose $-\pi/2 \le \theta < 0$ such that $\sin\theta = x$. Thus this way of choosing θ such that $\sin\theta = x$ for $-1 \le x < 1$ has no ambiguity. Such a value of the inverse circular function is called its *principal value*, though we could choose another set of values with equal ease.

Similar problems arise in the context of a function $f : \mathbb{R} \to \mathbb{R}$ defined by $f(x) = x^2$. The function f is neither one-to-one nor onto. But if we take $f : \mathbb{R} \to \mathbb{R}_+$ as $f(x) = x^2$, then f is onto. We would like to define $f^{-1} : \mathbb{R}_+ \to \mathbb{R}$ as a function. In order that we are able to define f^{-1} as a function we must agree, once and for all, the sign of $f^{-1}(x)$. Indeed, since $f(-1) = f(1) = 1$, which one would we call $f^{-1}(1)$? In fact, $f^{-1}(x)$ is what we would like to denote by $\sqrt{x}$. But we must decide if we are taking the positive value or the negative value. Once we decide that, f^{-1} would become a function.

Exercises

(i) Draw the graph of $f : \mathbb{R} \to \mathbb{R}$ defined by $f(x) = [x] =$ the greatest integer less than or equal to x.

(ii) Suppose $\Phi : \mathbb{R} \to \mathbb{R}$ is defined to be $\Phi(t) = |t|$ for $-1 \leq t \leq 1$ and $\Phi(t+2) = \Phi(t)$ for $t \in \mathbb{R}$. Draw its graph.

(iii) Let $f(x) =$ distance of x from the nearest integer. Draw the graph of f.

(iv) Draw the graph of $f : \mathbb{R} \to \mathbb{R}$ defined by $f(x) = x - [x]$.

(v) Let $f : \mathbb{R} \to \mathbb{R}$ be defined by

$$f(x) = \begin{cases} 1 & \text{for } 2n < x \leq 2n+1, (n \in \mathbb{Z}) \\ 0 & \text{for } 2n+1 < x \leq 2n+2. \end{cases}$$

Draw its graph.

Suggested Reading

1. *Calculus*, Vol. I by T. M. Apostol, Wiley Eastern, New Delhi 1980 (good Introduction).

2. *The Problems of Mathematics* by I. Stewart, Oxford University Press, New York 1992 (for general reading).

Chapter 2

Real Numbers, Sequences and Series

"The northern ocean is beautiful" said the Orc, "and beautiful is the delicate intricacy of the snowflake before it melts and perishes, but such beauties are nothing to him who delights in numbers, spurning alike the wild irrationality of life and the baffling complexity of nature's laws."

J.L. Synge

2.1 Natural Numbers

Natural numbers are perhaps the earliest mathematical notion of man. The difference between a pack of wolves and one wolf, a swarm of bees and one bee, a school of fish and a single fish, a heap of stones and a single stone, etc., may have led primitive man to systematize the idea of numbers. The fact that in Sanskrit (and in some other ancient Indo-European languages) there are usages where one object, two objects and many objects are distinguished, e.g.

नरः , नरौ , नराः

indicates man's attempt to distinguish between numbers. Heaps of stones found in caves, where early men lived, and the discovery of animal bones on which notches have been cut in regular series of five are indicative of devices by the human beings of yore to keep count of their livestock. That is how they understood plurality.

But what do *we* understand by numbers? Is number the same thing as plurality? We often use expressions like "there are *five* fingers on my hand", "Pandavas were *five* brothers", "there are *five* books on this table", "there are *five* mangoes in the basket", etc. Observe what we are talking about. In the first case we were talking about a set of fingers on my hand, in the next about a set of brothers, next about a set of books on a table, next about a set of mangoes. But to each of these sets we attached the common attribute "five". Why was that? There must have been something common in all these sets to justify our attaching the common attribute. The commonality between them is that the elements of one set can be put in one-to-one correspondence with the elements of the other sets under discussion.

We say that two sets A and B are *"equinumerous"* (or of the same cardinality) if there is a map

$$f : A \to B$$

which is one-to-one and onto, i.e., a bijection. One sees that if A and B are equinumerous and the sets B and C are equinumerous, then so are A and C. One can easily observe that being *equinumerous* is an equivalence relation among sets. For example, see Figure 2.1.

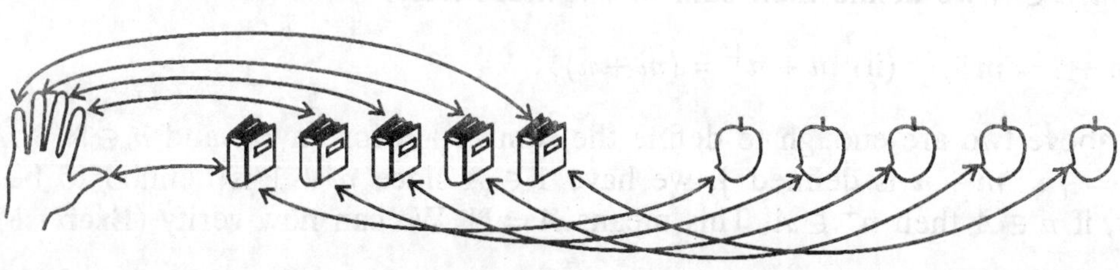

Figure 2.1

This relation decomposes any class of sets into disjoint equivalence classes of sets. Two sets are in the same equivalence class if they are equinumerous. Each equivalence class is characterized by a common property of its members, i.e., any two of them are equinumerous. This characterizing property is what we would like to call the *number* of elements of the set belonging to the equivalence class. Since equivalence classes, into which we have classified the sets, are disjoint, the characterising property associated with disjoint classes are distinct. "Commonness", "common property" or "characteristics" is too imprecise an idea to be handled easily; some authors have defined the number of elements of a set as the equinumerous class to which it belongs. Thus the number "two" would mean the class of *all pairs*, the number "three" the class of all *triplets*, $\cdots$ and so on. But this too has some foundational problems which are hard to circumvent. We do not go into the details of such things here.

Instead, we adopt the axioms devised by Peano for the natural numbers $\mathbb{N}$ with the following properties.

1. There is a map (called the *successor map*) from $\mathbb{N}$ to $\mathbb{N}$, sending an element n

to n^+ (called the *successor* of n), which is injective, i.e., $n \neq m \Rightarrow n^+ \neq m^+$.

2. There is an element 1 that is not a successor of any element (i.e., 1 is not in the image of the successor map, i.e., $1 \neq n^+$, for all $n \in \mathbb{N}$). Also, every other element of $\mathbb{N}$ is the successor of some element of $\mathbb{N}$.
3. Suppose that $A \subseteq \mathbb{N}$ and that (a) $1 \in A$, (b) $n^+ \in A$ whenever $n \in A$. Then $A = \mathbb{N}$, i.e., the only subset of $\mathbb{N}$ which contains 1 and the successor of each of its elements is $\mathbb{N}$ itself.

The third axiom is known as the *Principle of Mathematical Induction*.

2.2 Addition and multiplication of natural numbers

For $m, n \in \mathbb{N}$ we define their sum $m + n$ inductively:

(i) $m + 1 = m^+$, (ii) $m + n^+ = (m + n)^+$.

The above two are enough to define the sum $m + n$ for any m and $n \in \mathbb{N}$. Indeed, if $A = \{n : m + n \text{ is defined }\}$, we have $1 \in A$ since $m + 1$ is defined to be m^+. Now, if $n \in A$ then $n^+ \in A$. This means $A = \mathbb{N}$. We can now verify (**Exercise**)

(i) $m + n = n + m$, (ii) $m + (n + p) = (m + n) + p$.

For $m, n \in \mathbb{N}$ we define $m.n$ as

(i) $m.1 = m$, (ii) $m.n^+ = m.n + m$.

It is easy to verify that (**Exercise**)

(iii) $m \cdot n = n \cdot m$ for all $m, n \in \mathbb{N}$,

(iv) $m \cdot (n \cdot p) = (m \cdot n) \cdot p$, for all $m, n, p \in \mathbb{N}$,

(v) $m \cdot (n + p) = m \cdot n + m \cdot p$, for all $m, n, p \in \mathbb{N}$.

For $m, n \in \mathbb{N}$, we say that m is less than n, written as $m < n$, if there exists $k \in \mathbb{N}$ such that $m + k = n$.

After we have defined natural numbers we are in a position to define what is called a *finite set*. A non-empty set is called a *finite set* if there is a natural number n such that there is a bijection

$$f : A \to \{1, 2, \cdots, n\}. \tag{2.1}$$

The empty set is taken to be finite by convention. A set which is *not finite* is called an *infinite* set. This means that for a set A to be infinite one has to show that we cannot construct a bijection f as in (2.1) above. This may be inconvenient to actually implement in practice. The following criterion, due to Dedekind, gives a more convenient way of deciding if a set is infinite or not.

THEOREM: *A set is an infinite set if and only if it has a proper subset to which it is equinumerous.*

A formal proof of this is beyond the scope of this book.

It is easily seen that $\mathbb{N}$ is equinumerous to $\mathbb{N} - \{1\}$. Indeed the bijection defined below shows this: $f(n) = n + 1$, i.e.,

$$\begin{array}{ccccccc} 1 & 2 & 3 & \cdots & n-1 & \cdots \\ \updownarrow & \updownarrow & \updownarrow & & \updownarrow & \\ 2 & 3 & 4 & \cdots & n & \cdots \end{array}$$

This bijection is interestingly put as follows: There was a hotel with a room corresponding to each natural number. That is to say there are rooms with numbers $1, 2, \cdots$ and so on. Once when all the rooms were occupied by guests, a new customer arrived and wanted a room badly. How did the manager manage to accomodate the new customer without evicting any of the already existing guests from the hotel?

The manager knew mathematics. He simply requested the customer of room number 1 to move to room number 2, customer of room number 2 to room number 3 and so on. Now each customer has been accomodated without duplication where as room number 1 is vacant. He can now accomodate the new customer in that room. Thus we have been able to show that $\mathbb{N}$ is equinumerous to a proper subset $\mathbb{N} - \{1\}$ of itself. Hence it must be an infinite set. A set is said to be countably infinite if it is equinumerous to $\mathbb{N}$.

Integers

We have already seen that we can add and multiply two elements of $\mathbb{N}$. We can also "subtract" an element of $\mathbb{N}$ from another provided the element from which we are subtracting is greater than or equal to the element we are subtracting. If $m, n \in \mathbb{N}, m > n$, then there is a unique number $p \in \mathbb{N}$ such that

$$m = n + p,$$

and we write $p = m - n$.

This operation is what is called *subtraction* as we had learnt in our elementary school. But, as we have seen, we can subtract a number only from larger numbers and not the other way round. This is the same situation as when I have Rs. 10,000/- in my account in a bank and I write a cheque for Rs. 9,000/-, the bank *honours* the cheque and pays the payee Rs. 9,000/- deducting it from my account leaving a *balance* of Rs. 1,000/-. On the other hand, if I write a cheque for Rs. 15,000/-, the bank will normally *dishonour* the cheque. But sometimes, depending on my credit worthiness, the bank obliges me by allowing an *overdraft*, i.e., the bank remembers that I owe the bank Rs. 5000/- and yet allows me to maintain my account. This is sometimes called an account with a "negative" balance.

Let us assume that the banker decides to designate deposits in natural numbers and withdrawals in bold face numbers. While bringing the account up-to-date, if the withdrawals total less than the deposits, then the balance is entered in natural numbers. If the withdrawals total more than the deposits, then the excess of the withdrawals over the deposits (what is called the overdraft) is entered in the balance column in bold face numbers, meaning that the account holder owes the amount, entered in bold face, to the bank. When he deposits some amount next time the balance is entered after deducting the overdraft from the deposit. Thus the banker introduces a new set of numbers which he writes in bold face. Similarly, he writes 0 when the balance is nil. But the rules of additon and subtraction are more general than those for the natural numbers: For example,

$$5 + \mathbf{4} = 1$$

meaning that if one has a balance of Rs.5/- and withdraws Rs.4/- then the balance of Re.1/- remains. Similarly, one writes

$$100 + \mathbf{150} = \mathbf{50}$$

implying that if one withdraws Rs. 150/- out of a total deposit of Rs. 100/-, then one is left with an overdraft of Rs. 50/-, or

$$\mathbf{250} + 300 = 50$$

meaning that if one has an overdraft of Rs. 250/- and deposits Rs. 300/-, then his net balance is Rs. 50/-. Also,

$$180 + 0 = 180, \qquad 70 + \mathbf{70} = 0$$

meaning that if the balance is 180 and nothing is deposited nor withdrawn, then the balance is still 180 and if the entire balance (of Rs. 70) is withdrawn, then the balance becomes nil.

Thus we observe that the banker, by augmenting the natural numbers with a new set of numbers which are written in bold face, is able to add and subtract at will without any restriction. The banker's numbers now consist of the natural numbers and a set of new numbers which he writes in bold face. So his number system now is

$$\{\cdots, \mathbf{3}, \mathbf{2}, \mathbf{1}, 0, 1, 2, 3, \cdots\}.$$

It should now be clear that the operation of subtraction which was earlier of restricted applicability can now be used in this new augmented number system by setting $0 - n = \mathbf{n}$, where $n \in \mathbb{N}$.

We may use the notation

$$\mathbf{1} = -1, \mathbf{2} = -2, \mathbf{3} = -3, \cdots,$$

where each bold face number is replaced by a number with a negative sign prefixed. Now our augmented number system is

$$\{\cdots, -3, -2, -1, 0, 1, 2, 3, \cdots\}.$$

We denote this changed set of numbers by $\mathbb{Z}$ ($\mathbb{Z}$ is the first letter of the German word *Zahlen* which means numbers). We have thus

$$\mathbb{Z} = \{\cdots, -3, -2, -1, 0, 1, 2, 3, \cdots\}.$$

Clearly, $\mathbb{N} \subset \mathbb{Z}$ and we would expect $\mathbb{Z}$ to be amenable to at least the operations we could perform on $\mathbb{N}$. Let us take the operation of addition first. For $m, n \in \mathbb{Z}$ there is a unique number written $m + n \in \mathbb{Z}$ such that

a) $m + n = n + m$ for all $m, n \in \mathbb{Z}$,

b) $m + (n + p) = (m + n) + p$ for all $m, n, p \in \mathbb{Z}$,

c) $m + 0 = m$ for all $m \in \mathbb{Z}$,

d) for every $m \in \mathbb{Z}$ there is a number $n \in \mathbb{Z}$ such that

$$m + n = 0.$$

One can show that this number n is unique and we write $n = -m$. We have automatically

$$m + n = m \Rightarrow n = 0.$$

This leads to an even more interesting property

$$-(-m) = m \text{ for all } m \in \mathbb{Z}.$$

Indeed, for every $m \in \mathbb{Z}$ there is a unique $-m \in \mathbb{Z}$ such that

$$m + (-m) = 0.$$

Again by the same argument there is $(-m) \in \mathbb{Z}$ such that

$$-m + (-(-m)) = 0.$$

Adding m to both the sides we get

$$m + (-m + (-(-m))) = m + 0 = m.$$

But by associativity we have

$$m + (-m + (-(-m))) = (m + (-m)) + (-(-m)).$$

This gives $0 + (-(-m)) = m$ which is the same as $-(-m) = m$.

Next, consider multiplication. For all $m, n \in \mathbb{Z}$ we should have a unique $m \cdot n \in \mathbb{Z}$ having the properties

1. $m \cdot n = n \cdot m$ for all $m, n \in \mathbb{Z}$,

2. $m \cdot (n \cdot p) = (m \cdot n) \cdot p$ for all $m, n, p \in \mathbb{Z}$.

Besides, where it involves addition and multiplication, the *distributive* rule

3. $m \cdot (n + p) = m \cdot n + m \cdot p$ for all $m, n, p \in \mathbb{Z}$,

should be satisfied.

We can now demonstrate

i) $m \cdot 0 = 0 \cdot m = 0$ for all $m \in \mathbb{Z}$,

ii) $(m + n) \cdot p = m \cdot p + n \cdot p$ for all $m, n, p \in \mathbb{Z}$,

iii) $m \cdot 1 = 1 \cdot m = m$ for all $m \in \mathbb{Z}$,

iv) $m \cdot (-n) = (-m) \cdot n = -(m \cdot n)$ for all $m, n \in \mathbb{Z}$,

v) $(-m) \cdot (-n) = m \cdot n$ for all $m, n \in \mathbb{Z}$.

Part (i) is clear because we have the relation $n + 0 = n$. Multiplying by $m \in \mathbb{Z}$, we get

$$(n + 0) \cdot m = n \cdot m + 0 \cdot m = n \cdot m \quad \text{implying the required result.}$$

ii) $(m + n) \cdot p = p \cdot (m + n) = p \cdot m + p \cdot n = m \cdot p + n \cdot p.$

iv) Since $m + (-m) = 0$, we have $m \cdot n + (-m) \cdot n = 0$ which leads to

$$(-m) \cdot n = -(m \cdot n).$$

Similarly, $n + (-n) = 0$ implies that $m \cdot (n + 1(-n)) = 0$, i.e., $m \cdot n + m(-n) = 0$. This gives

$$m \cdot (-n) = -(m \cdot n).$$

This also leads to

$$(-m) \cdot (-n) = -((-m) \cdot n) = -(-(m \cdot n)) = m \cdot n$$

which is (v). This demonstrates how to derive this relation (which must be well known from high school) from the defining properties of $\mathbb{Z}$.

iii) $m \cdot 1 = m$ for $m \in \mathbb{N}$. If $m \notin \mathbb{N}$ we have $-m \in \mathbb{N}$. Hence

$$(-m) \cdot 1 = -(m \cdot 1) = -m.$$

Thus we have

$$m \cdot 1 = m.$$

The elements of $\mathbb{Z}$ are called *integers* or *whole numbers*.

Pictorial Representation of Integers $\mathbb{Z}$

Take a straight line and fix a point on it. Call that point 0. Measure a fixed distance to the right of 0 and mark the point with that distance as 1. Now mark a point to the right of point 1 at the same distance from 1 as the distance of 1 from 0 and call that point 2. Continue this process to get all the *natural numbers* (Figure 2.2).

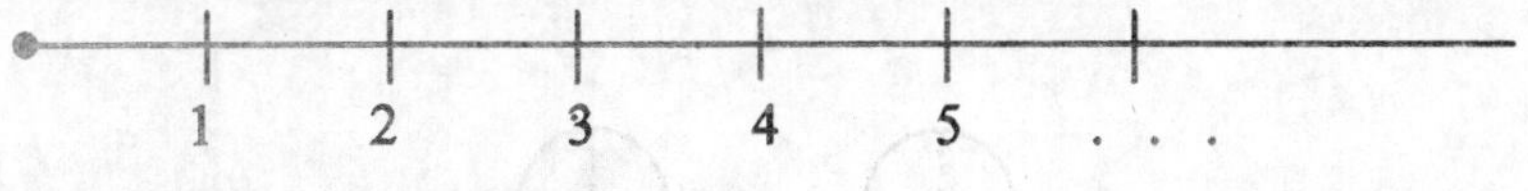

Figure 2.2

Take a point to the left of 0 at as much distance as 1 is to the right of 0. Call this point -1. Similarly, the point to the left of -1 at a distance equal to the distance between 0 and -1 is called -2. We can continue this process to get all the *negative* integers (Figure 2.3).

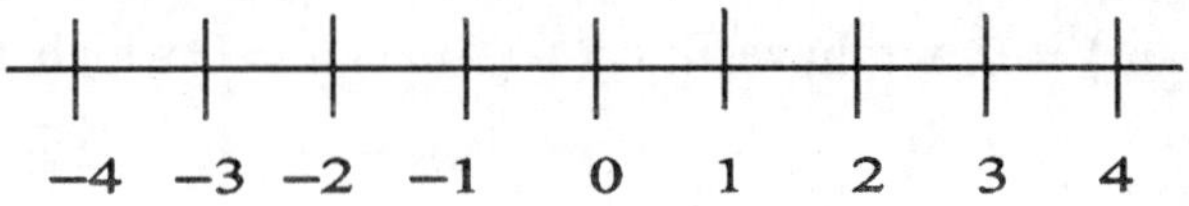

Figure 2.3

On the line with all the integers marked, addition or subtraction can be thought of as going forward or backward from a given position by a certain number of steps. For example, 3 + 4 would mean four steps to the right beginning with 3, where 3 – 4 would mean four steps to the left starting from this point representing 3. So we come to one step left of 0 giving the result –1. The subtraction of a negative number can be interpreted again as an advance to the right.

2.3 *Rational Numbers*

On $\mathbb{Z}$ we now have unlimited subtraction. Can we divide in $\mathbb{Z}$? In a limited sense, yes. Division is essentially an inverse problem. That is to say, $m \div n$ would stand for a number which when multiplied by n gives m. As long as we are in $\mathbb{Z}$ the question does not always have an answer. For example we know that 5 multiplied by 2 gives 10 as the product, hence 10 divided by 5 gives 2 as the *quotient.* What if we asked which number multiplied by 2 gives us 5? We know there is no integer which when multiplied by 2 gives 5. Such questions frequently arise when we have to distribute m things equally among n persons. If m is not an integral multiple of n, then $\mathbb{Z}$ becomes inadequate for such distribution.

If 10 cakes are to be shared equally among 5 children, then each child gets 2 cakes. But how do we proceed if the same 10 cakes are to be shared equally among 4 children? One possibility is to give 2 cakes to each of the 4 children and then divide each of the remaining 2 cakes into two equal halves so that there are 4 halves. Each child can now be given half a cake. Pictorially, they are represented as in Figure 2.4.

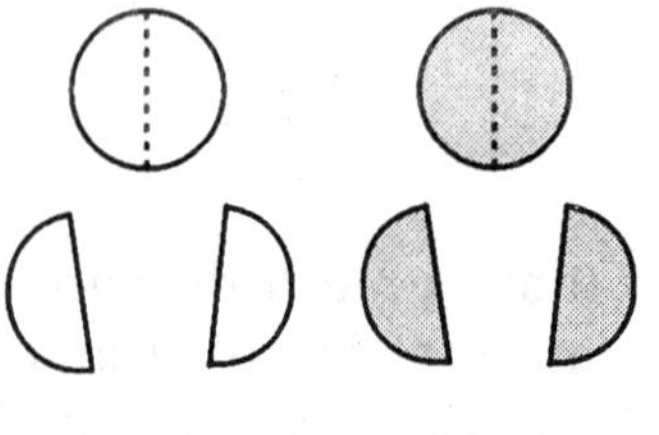

Figure 2.4

What we are doing is introducing a new kind of numbers called "fractions", meaning a "part" of the whole number. Each half of the cake is represented by $\frac{1}{2}$. Suppose we had 3 cakes to be shared equally by 5 children. What we do is to divide each cake into 5 equal parts. So we have in all 15 similar pieces of cake. Now we can distribute these 15 pieces among 5 children, each getting a share of 3 pieces. Each child's share is represented by $\frac{3}{5}$. Suppose we divided each of the cakes into 10 equal parts instead, then we would have thirty equal pieces of cake. When shared equally by 5 children, each child gets a share of 6 pieces. Each child's share can be represented by $\frac{6}{10}$. The share of each child in both ways of division ought to be the same, as in both cases each child has an equal share and nothing of the original three cakes remains. This is what would amount to saying that $\frac{3}{5}$ and $\frac{6}{10}$ represent the same number. Geometrically, they look identical as illustrated in Figure 2.5.

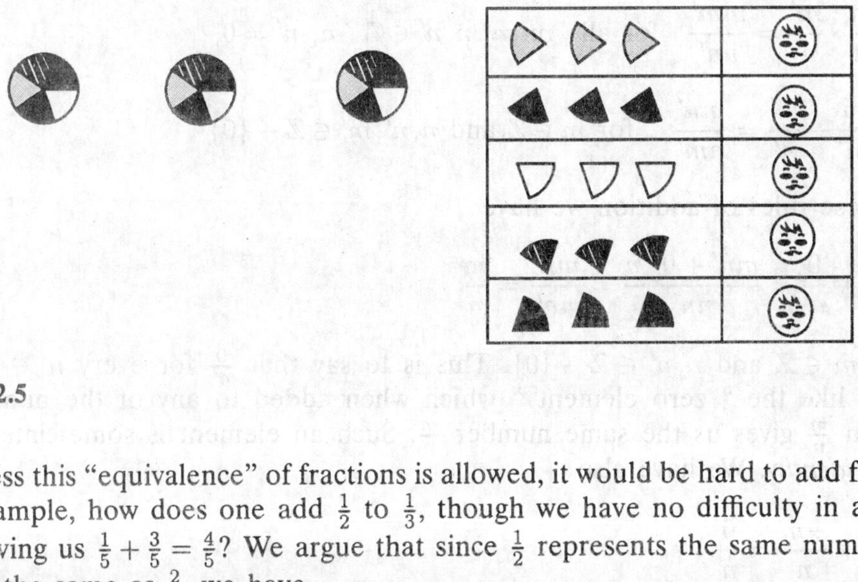

Figure 2.5

Unless this "equivalence" of fractions is allowed, it would be hard to add fractions. For example, how does one add $\frac{1}{2}$ to $\frac{1}{3}$, though we have no difficulty in adding $\frac{1}{5}$ to $\frac{3}{5}$ giving us $\frac{1}{5}+\frac{3}{5}=\frac{4}{5}$? We argue that since $\frac{1}{2}$ represents the same number as $\frac{3}{6}$ and $\frac{1}{3}$, the same as $\frac{2}{6}$, we have

$$\frac{1}{2}+\frac{1}{3}=\frac{5}{6}.$$

This means that if somebody had three parts out of six parts, and next two parts out of six parts, then he has in all five parts out of six. We can perform subtraction similarly. Multiplication and division can be defined as we did in our elementary school. This presupposes that a fraction represents measurement of some physical body like cake, stick, pole, etc. But we should have a definition which should capture the essence of the process of "dividing a certain physical object like a cake into 5 equal parts". To formalize our construction of fractions we proceed as follows:

Consider the Cartesian product $\mathbb{Z} \times (\mathbb{Z} - \{0\}) = \{(m, n) : m, n \in \mathbb{Z}, n \neq 0\}$ and define the relation $\sim$ in $\mathbb{Z} \times (\mathbb{Z} - \{0\})$ by $(m, n) \sim (m', n')$ if $mn' = m'n$. It is easy to see that $\sim$ is an equivalence relation in $\mathbb{Z} \times (\mathbb{Z} - \{0\})$.

Now we can decompose the set $\mathbb{Z} \times (\mathbb{Z} - \{0\})$ into disjoint equivalence classes. Let us denote the equivalence class containing (m, n) by $\frac{m}{n}$.

Suppose (m', n') is another element of the same equivalence class. Then we ought to denote it also by $\frac{m'}{n'}$. But since $(m, n) \sim (m', n')$ we have $mn' = m'n$. So we have $\frac{m}{n} = \frac{m'}{n'}$ if $mn' = m'n$. We can now define addition, subtraction, multiplication and division among new numbers denoted by $\frac{m}{n}$ in the following way:

(i) $\dfrac{m}{n} + \dfrac{m'}{n'} = \dfrac{mn' + m'n}{nn'}$ for all $m, n, m', n' \in \mathbb{Z}$, $n, n' \neq 0$,

(ii) $\dfrac{m}{n} \cdot \dfrac{m'}{n'} = \dfrac{mm'}{nn'}$ for all $m, m', n, n' \in \mathbb{Z}$ $n, n' \neq 0$,

(iii) $\dfrac{m}{n} \div \dfrac{m'}{n'} = \dfrac{mn'}{nm'}$ for $m \in \mathbb{Z}$ and $n, n', m' \in \mathbb{Z} - \{0\}$.

With these rules of addition we have

$$\frac{m}{n} + \frac{0}{n'} = \frac{mn' + 0 \cdot n}{nn'} = \frac{mn'}{nn'} = \frac{m}{n}$$

for all $m \in \mathbb{Z}$ and $n, n' \in \mathbb{Z} - \{0\}$. This is to say that $\frac{0}{n'}$ for every $n' \in \mathbb{Z} - \{0\}$ behaves like the " zero element " which when added to any of the numbers of the form $\frac{m}{n}$ gives us the same number $\frac{m}{n}$. Such an element is some times called *additive identity*. We have also

$$\frac{m}{n} + \frac{-m}{n} = \frac{0}{n}.$$

This tells us that for every element $\frac{m}{n}$, we have an element of the same type, which when added to it gives the zero element. This is what we usually understand as the "negative" of the number $\frac{m}{n}$ which is sometimes called the *additive inverse* of $\frac{m}{n}$. Again, the rules of multiplication tell us that

$$\frac{m}{n} \cdot \frac{n'}{n'} = \frac{mn'}{nn'} = \frac{m}{n} \quad \text{for all } m, n, n' \in \mathbb{Z} \text{ and } n \neq 0, n' \neq 0.$$

That means that $\frac{n'}{n'}$ acts the way the *multiplicative identity* like 1 behaves in $\mathbb{Z}$ (which when multiplied by any number gives us back the same number). For simplicity of notation we denote $\frac{0}{n}$ by 0 (note that $\frac{0}{n} \sim \frac{0}{n'}$) and $\frac{n}{n}$ by 1 so that we may identify $\frac{mn}{n}$ with $m \in \mathbb{Z}$. Since $\frac{m}{n} \cdot \frac{n}{m} = \frac{m \cdot n}{m \cdot n} = 1$ when $m, n \in \mathbb{Z}$ and $m \neq 0, n \neq 0$, we conclude that every non-zero element of the form $\frac{m}{n}$ has a *multiplicative inverse.*

We call the new numbers $\frac{m}{n}$ *rational numbers* which is in fact an extension of $\mathbb{Z}$. We denote the set of rational numbers by

$$\mathbb{Q} = \left\{\frac{m}{n} : m, n \in \mathbb{Z}, n \neq 0\right\}.$$

In $\mathbb{Q}$ we can add, subtract and multiply at will. We can also verify easily (**Exercise**):

(i) $\alpha + \beta = \beta + \alpha$ for all $\alpha, \beta \in \mathbb{Q}$,

(ii) $\alpha + (\beta + \gamma) = (\alpha + \beta) + \gamma$ for all $\alpha, \beta, \gamma \in \mathbb{Q}$,

(iii) $\alpha + 0 = \alpha$ for all $\alpha \in \mathbb{Q}$,

(iv) for every $\alpha \in \mathbb{Q}$, there is a unique $\beta \in \mathbb{Q}$ such that $\alpha + \beta = 0$.

We call this β the negative of α, denote it by $-\alpha$ and write $\alpha + (-\alpha) = 0$.

(v) $\alpha \cdot \beta = \beta \cdot \alpha$ for all $\alpha, \beta \in \mathbb{Q}$,

(vi) $\alpha \cdot (\beta \cdot \gamma) = (\alpha \cdot \beta) \cdot \gamma$ for all α, β, $\gamma \in \mathbb{Q}$,

(vii) $\alpha \cdot 1 = \alpha$ for all $\alpha \in Q$.

(viii) $\alpha \cdot (\beta + \gamma) = \alpha \cdot \beta + \alpha \cdot \gamma$ for all α, β, $\gamma \in \mathbb{Q}$,

(ix) for every $\alpha \in \mathbb{Q}$, $\alpha \neq 0$, there is a unique $\beta \in \mathbb{Q}$ such that $\alpha \cdot \beta = 1$.
We denote $\beta = \frac{1}{\alpha}$ or α^{-1}.

Given $\alpha \in \mathbb{Q}$ and $\beta \in \mathbb{Q} - 0$, we define α divided by β, written as $\frac{\alpha}{\beta}$ by $\alpha \cdot \beta^{-1}$. Note that for $\beta \neq 0$, $\frac{\alpha}{\beta} = \gamma \Leftrightarrow \alpha = \beta\gamma$.

Remark: Note that if $\alpha \neq 0$ and $\beta = 0$, then there does not exist a γ such that $\alpha = \beta\gamma$. Thus in this case, $\frac{\alpha}{\beta}$ cannot be defined. Also, when $\alpha = 0$ and $\beta = 0$, any choice of γ will satisfy $\alpha = \beta\gamma$. Hence again, $\frac{\alpha}{\beta}$ cannot be defined. So, division by zero is not defined.

Remark: Any set F with two operations $(+, \cdot)$ satisfying all the nine properties listed above is called a *field*. This means that $\mathbb{Q}$ is a field. For a detailed discussion on fields, see Chapter 7.

The set $\mathbb{Q}$ is called the set of rational numbers. No doubt $\mathbb{Q}$ is a field where usual arithmetical operations can be performed, but it has an additional feature that is the *order relation*. That is to say, given two rational numbers we can decide, if they are not equal, which is the larger of the two. To describe the modus operandi of this we proceed as follows:

Let $\mathbb{Q}_+ = \{\frac{m}{n} : m, n > 0\}$, the subset of positive rational numbers, then $\mathbb{Q}_+$ is a subset of $\mathbb{Q}$ having the properties

i) for $\alpha \in \mathbb{Q}$ either $\alpha \in \mathbb{Q}_+$ or $\alpha = 0$, or $-\alpha \in \mathbb{Q}_+$.

In other words, $\mathbb{Q}_+ \cup \{0\} \cup \{-\alpha : \alpha \in \mathbb{Q}_+\} = \mathbb{Q}$.

ii) $\alpha, \beta \in \mathbb{Q}_+$ implies that $\alpha + \beta,\ \alpha\beta \in \mathbb{Q}_+$.

This is sometimes paraphrased by saying that $\mathbb{Q}_+$ is closed under addition and multiplication. We define an order relation in $\mathbb{Q}$ with the help of $\mathbb{Q}_+$ by saying that for $\alpha, \beta \in \mathbb{Q}$, $\alpha \neq \beta$,

$$\alpha < \beta \text{ if } \beta - \alpha \in \mathbb{Q}_+ \text{ or } \beta < \alpha \text{ if } \alpha - \beta \in \mathbb{Q}_+.$$

This is meaningful because either $\beta - \alpha \in \mathbb{Q}_+$ or $-(\beta - \alpha) \in \mathbb{Q}_+$ or $\beta - \alpha = 0$.

i) Given $\alpha, \beta \in \mathbb{Q}$ and $\gamma \in \mathbb{Q}_+$, we have $\alpha < \beta$ implies $\alpha\gamma < \beta\gamma$; indeed, this is obvious because we have $\beta - \alpha \in \mathbb{Q}_+$ and hence $(\beta - \alpha)\gamma \in \mathbb{Q}_+$.

ii) $0 < \alpha, \beta$ and $\alpha < \beta \Rightarrow \frac{1}{\beta} < \frac{1}{\alpha}$.

iii) $\alpha < \beta \ \Rightarrow \ -\beta < -\alpha$.

iv) $\alpha < \beta,\ \gamma < 0 \ \Rightarrow \ \beta\gamma < \alpha\gamma$.

v) Given $\alpha, \beta \in \mathbb{Q}$ and $0 < \alpha$, there is an $n \in \mathbb{N}$ such that $\beta < n\alpha$.

For a proof of this, first note that if $\beta < \alpha$ then there is nothing to prove (we might take $n = 1$). If on the other hand $\alpha < \beta$, set $\alpha = \frac{p}{q}$and $\beta = \frac{p'}{q'}$, $p, q, p', q' \in \mathbb{N} - \{0\}$. This is true because $0 < \alpha < \beta$.

Now $\alpha q = p$. Let $\frac{p'}{q'} = n_1 + \frac{r'}{q'}$, where $n_1 \in \mathbb{N} \cup \{0\}$ and $0 \leq r' < q$. Also, we can always find an $m \in \mathbb{N}$ such that $mp > n_1 + \frac{r'}{q'}$ (since $0 \leq \frac{r'}{q'} < 1$). Finally we set $qm = n$ so that we have

$$n\alpha = mq\alpha = mp > \frac{p'}{q'} = \beta.$$

This last property is called the *Archimedean property* of $\mathbb{Q}$. (This was used by Archimedes in his works but he credited Eudoxus for this).

vi) Given $\alpha, \beta \in \mathbb{Q}$, $\alpha < \beta$ there is a $\gamma \in \mathbb{Q}$ such that $\alpha < \gamma < \beta$. For example, take $\gamma = \frac{\alpha+\beta}{2}$.

vii) $\alpha \leq \beta \Rightarrow \alpha + \gamma \leq \beta + \gamma$ for all $\gamma \in \mathbb{Q}$.

It can be verified that $0 \cdot \alpha = 0$ and $(-\alpha) \cdot (-\beta) = \alpha \cdot \beta$ for all $\alpha, \beta \in \mathbb{Q}$.

Exercises

Prove the following:

(i) $-1 < 0$ and

(ii) $\alpha^2 > 0$ for all $\alpha \in \mathbb{Q}, \alpha \neq 0$.

Representation of $\mathbb{Q}$ on the Line

We have seen earlier that every element of $\mathbb{Z}$ can be represented on a straight line with a certain point representing 0. Positive integers are represented by points to the right of this marked point at equal lengths, and negative integers are similarly represented by points to the left of this.

We can use the same straight line to represent the newly constructed rational numbers as follows:

We know that we can divide a line segment, using a straight edge and compass, into q equal parts. So the line segment between 0 to 1 can be divided into q equal parts for every positive integer q (Figure 2.6).

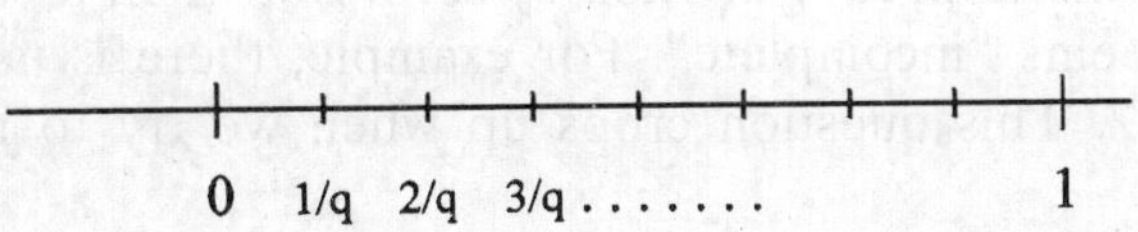

Figure 2.6

The points of division will now represent the numbers $\frac{1}{q}, \frac{2}{q}, \cdots, \frac{q-1}{q}$. It is now clear that any rational number of the form $\frac{p}{q}$ can be represented on the straight line. We have seen that between any two distinct rational numbers r and r', *no matter how close they are*, we can always find another rational number between them, e.g., $\frac{r+r'}{2}$. Geometrically, this means that between any two distinct points on the above line representing rational numbers, there is a point between them representing a rational number.

One can ask if every point on the line corresponds to a rational number. We have a point on the line which corresponds to the length of the diagonal of a square of side 1. But we shall show in the next section that the length of the diagonal of the unit square does not correspond to a rational number.

To prepare the ground for such discussions, let us start with a few definitions: A subset $A \subset \mathbb{Q}$ is said to be *bounded below* if there exists a rational number α such that

$$\alpha \leq \beta \quad \text{for every} \quad \beta \in A,$$

and α is called a *lower bound* of A. Similarly, $A \subset \mathbb{Q}$ is said to be *bounded above* if there is a $\gamma \in \mathbb{Q}$ such that $\beta \leq \gamma$ for every $\beta \in A$, and γ is called an *upper bound* of A. A set which is bounded above and also bounded below is called a *bounded set*. Observe that if a set has a lower bound α then it has many more like $\alpha - 1, \alpha - 2, \cdots$. Similarly, a set that is bounded above has many upper bounds. If there is a least element of all these upper bounds, then we call it the *least upper bound* of the set. But it is not true that every set of rational numbers which is bounded above has a least upper bound. We can say similar things about sets which are bounded below. Some sets may have least upper bound. For example, the set of negative rational numbers is certainly bounded above and it is easy to see that 0 is its least upper bound.

2.4 Real Numbers

Our scheme of numbers has so far been extended to rational numbers so that we could not only add and subtract without any hindrance, but could also multiply and divide without any restriction, save division by zero. But when it comes to extraction of square roots, it seems "incomplete". For example, there is no rational number whose square gives 2. This question crops up when we try to relate numbers to geometry.

We have seen before that we can always represent a rational number on a line. Now suppose we have a square with sides of unit length. By the theorem of Pythagoras, the square of the length of the diagonal is equal to $1^2 + 1^2 = 2$. We can also see that the length of the diagonal can be geometrically represented on the line. Now we ask here, does there exist a rational number whose square is 2 or, equivalently, does the equation $x^2 = 2$ have a solution in rational numbers? This amounts to asking if we can find two non-zero integers p and q such that

$$p^2 = 2q^2. \tag{2.2}$$

We can demonstrate that there are no such integers. Suppose there were two such integers p and q satisfying (2.2). We may assume, without loss of generality, that p and q have no common factors. If there were any, we could cancel them from both the sides of (2.2) till there were no common factors. Now as there are no common factors between p and q, for (2.2) to old, 2 must be a factor of p^2, hence of p. So we may write

$$p = 2m. \tag{2.3}$$

Substituting this in (2.2) we get

$$2m^2 = q^2 \tag{2.4}$$

As before, we conclude that for (2.4) to hold, 2 must divide q giving $q = 2n$ for some integer n. This contradicts our assumption that p and q have no common factors. This only proves that (2.2) has no solution in integers (except $p = 0, q = 0$). Thus the length of the diagonal of the unit square, though it has a point representing it on the line, does not correspond to a rational number. This shows that $\mathbb{Q}$ is not large enough to accommodate a number such as the length of the diagonal of the unit square. So we must extend $\mathbb{Q}$. We could simply include the new number which we write as $\sqrt{2}$ in the scheme along with $\mathbb{Q}$. But then this *ad hoc* extension may not stand upto all our demands to include newer and newer numbers which arise out of different algebraic equations, e.g. $\sqrt[3]{6}$, etc. Even if we somehow include these numbers in our scheme, we must know how to perform arithmetical operations like addition, subtraction, multiplication and division in it. Also, how does one decide which of a given pair of new numbers is smaller? There are many ways of extending $\mathbb{Q}$ to accommodate all these new numbers. We illustrate one of the simplest ways of doing this.

Let $E = \{\alpha \in \mathbb{Q} : \alpha^2 < 2\}$. It is clear that E has an upper bound, for example, 2. Next we note that if $y \in E$ is an upper bound of E, then there exists $n \in \mathbb{N}$ with $n(2 - y^2) > 1 + 2y$ (we can always choose such an n because of the Archimedean property of $\mathbb{Q}$). Then one has $2-(y+\frac{1}{n})^2 = (2-y^2)-\frac{1}{n}(2y+\frac{1}{n}) \geq (2-y^2)-\frac{2y+1}{n} > 0$, showing that $y^2 < (y+\frac{1}{n})^2 < 2$ and thus no upper bound of E can be in E. Let $x \in \mathbb{Q}$ be one such upper bound of E, then so is $\frac{x}{2} + \frac{1}{x}$ since $(\frac{x}{2} + \frac{1}{x})^2 - 2 = (\frac{x}{2} - \frac{1}{x})^2 > 0$. On the other hand, $x - (\frac{x}{2} + \frac{1}{x}) = \frac{x^2-2}{2x} > 0$ since $x \notin E$. Therefore the E has no least upper bound in $\mathbb{Q}$.

This tells us what to do. We now extend the field of rationals to a larger field containing it which has all the properties of $\mathbb{Q}$ along with an additional property, called the *least upper bound property* and abreviated as *lub* property, namely: *every set that is bounded above has a least upper bound.*

More precisely, we postulate that there is a field $(\mathbb{R}, +, \cdot)$, $\mathbb{R}$ containing $\mathbb{Q}$ and satisfying all the properties of $\mathbb{Q}$ listed above and an additional one:

The *least upper bound property* (*lub*): If $A \subseteq \mathbb{R}$ is bounded above, then A admits a least upper bound, i.e., (1) there is a $\gamma \in \mathbb{R}$ such that $\alpha \leq \gamma$ for every $\alpha \in A$ and (2) for every $\gamma' < \gamma$, there is a $\beta \in A$ such that $\gamma' < \beta$.

Remark: As we constructed $\mathbb{Q}$ out of $\mathbb{Z}$, there are ways of constructing $\mathbb{R}$ (with the properties stated) out of $\mathbb{Q}$. However, an explicit construction of $\mathbb{R}$ from $\mathbb{Q}$ is beyond the scope of this book. Nevertheless, we are going to use the listed properties of $\mathbb{R}$ in all subsequent discussions and deduce many interesting consequences.

Since $\mathbb{R}$ has all the properties of $\mathbb{Q}$ including the order property, we write (as for $\mathbb{Q}$),

$$\mathbb{R}_+ = \{\alpha \in \mathbb{R} | \alpha > 0\}$$

and this has the same properties as $\mathbb{Q}_+$. Now we use the same notation like $\leq, <, \geq, >$ for elements of $\mathbb{R}$, exactly as we did for the elements of $\mathbb{Q}$. As before, we write $\beta > \alpha$ or $\beta \geq \alpha$ according as $\alpha < \beta$ or $\alpha \leq \beta$.

Clearly $1,\ 1+1,\ 1+1+1, \cdots$ belong to $\mathbb{R}$ and represent the natural numbers. So one sets up a one-to-one correspondence between $n \in \mathbb{N}$ and $1 + 1 + \cdots + 1$(n times) to conclude that $\mathbb{N} \subseteq \mathbb{R}$. Writing $-n$ as the additive inverse of $n \in \mathbb{N}$, of course $0 \in \mathbb{R}$, we see that $\mathbb{Z} \subseteq \mathbb{R}$. Since $\mathbb{R}$ is a field, for $m, n \in \mathbb{Z},\ n \neq 0$, we must have

$$m.n^{-1} \in \mathbb{R},$$

where n^{-1} is the multiplicative inverse of n. We agree to write $m \cdot n^{-1} = \frac{m}{n}$. This way we have $\mathbb{Q} \subseteq \mathbb{R}$. We have thus effectively extended $\mathbb{Q}$ to $\mathbb{R}$.

Exercises

Using the properties of real numbers, show that

(i) $\alpha < 0$ or $\alpha = 0$ or $\alpha > 0$,

(ii) if $\alpha \neq 0,\ (-\alpha)^2 = (\alpha)^2 > 0$,

(iii) if $0 < a \leq b$, then $\frac{1}{b} \leq \frac{1}{a}$,

(iv) if $0 \leq a, b$, then $(1-a)(1-b) \geq 1 - a - b$,

(v) if $0 < a < 1$, then $a^n < 1$ for any positive integer n,

(vi) for $0 < a, b$, $\quad a^n < b^n$ implies $a < b$,

(vii) if $0 \leq a$, then $(1+a)^n \geq 1 + na$, and $(1-a)^n \geq 1 - na$ and

(viii) if $0 < a < \frac{1}{n}$, then $(1+a)^n < \frac{1}{(1-a)^n} < \frac{1}{1-na}$.

(ix) every set $A \subseteq \mathbb{R}$ that is bounded below admits a greatest lower bound.

For $x \in \mathbb{R}$, we define the absolute value of x (also called modulus of x), denoted by $|x|$ as

$$|x| = \begin{cases} x & \text{if } x \geq 0 \\ -x & \text{if } x < 0 \end{cases}.$$

We observe that $|x|$ is always non-negative. It is easy to check that for $x, y \in \mathbb{R}$,

$$|x+y| \leq |x| + |y|.$$

Here, if x and y are of the same sign, then equality holds.

We will now show that there is an $\alpha > 0$ in $\mathbb{R}$ such that $\alpha^2 = 2$, and this α is denoted by $\sqrt{2}$. In fact, we will prove much more.

THEOREM: *Given $x \in \mathbb{R}_+$ and $n \in \mathbb{N}$ we can find $y \in \mathbb{R}$ such that*

$$x = y^n.$$

PROOF: Let $A = \{u \in \mathbb{R}_+ : u^n < x\}$. If $x < 1$, then $x^n < x$ and hence $x \in A$. On the other hand, if $x \geq 1$, then $1/2 \in A$. Thus A is non-empty. Next, observe that if $v^n > x$, then v is an upper bound of A. In particular, $(1+x)$ is an upper bound of A.

By the least upper bound property, A admits a least upper bound. Let us denote it by y. We will rule out the possibilities $y^n < x$ as well as $y^n > x$, implying that $y^n = x$.

If $y^n < x$, let $a = \frac{x-y^n}{nx}$ and $z = y(1+a)$. It can be checked that $z^n < x$ so that $z \in A$. But $y < z$, contradicting the fact that y is the least upper bound of A (we have used the inequalities 7 and 8 in the **Exercise** above).

On the other hand, if $y^n > x$, let $b = \frac{y^n - x}{ny^n}$ and $w = y(1-b)$. Again, it can be verified that $w^n > x$ and hence w is an upper bound of A. But $w < y$, contradicting the fact that y is the least upper bound of A. Hence $y^n = x$. ■

In particular, we see that there is an element $\alpha \in \mathbb{R}$ such that $\alpha^2 = 2$ and hence also $(-\alpha)^2 = 2$ which means that the equation $x^2 = 2$ has two solutions. The positive one of these two solutions is $\sqrt{2}$. In fact, the above theorem has guaranted us extraction of the square root, cube root, even nth root of any positive number. You could ask at this stage, if this guarantees us extraction of square root of a negative number. The answer is clearly *no*. Indeed, we have

$$x^2 \geq 0, \quad \text{for } x \in \mathbb{R}.$$

Remark: We can further extend $\mathbb{R}$ to include "numbers" whose squares are negative. This you already know leads to the idea of *complex numbers*.

We have shown that $\mathbb{Q}$ is a subset of $\mathbb{R}$. One can also show that between *any two distinct real numbers there is a rational number*. This naturally leads to the *decimal representation of real numbers*: Given any real number x and any $q \in N$, we can get a unique $a_0 \in \mathbb{Z}$ and unique $a_1, a_2, \cdots, a_q \in \mathbb{N}$ such that $0 \leq a_1, \cdots, a_q \leq 9$ and

$$\left|x - \left(a_0 + \frac{a_1}{10} + \frac{a_2}{10^2} + \cdots + \frac{q_q}{10^q}\right)\right| < \frac{1}{10^q}.$$

You are invited to try to prove this familiar *decimal representation*.

If we have a terminating decimal representation of a real number, then surely it is rational. But we know that rationals like $\frac{1}{3}$, $\frac{1}{7}$, $\frac{1}{11}$, do not have a terminating decimal expansion.

$$\frac{1}{3} = 0.3333\cdots, \quad \frac{1}{7} = 0.142857\cdots, \text{ etc.}$$

It is clear that the decimal representation of $\sqrt{2}$ cannot terminate as it is not rational. There are many elements of $\mathbb{R}$ which are not in $\mathbb{Q}$. Any element of $\mathbb{R}$ which is not in $\mathbb{Q}$ is called an *irrational number*, and irrational numbers cannot have terminating decimal representations.

Exercise: Discover (and justify) an essential difference between the decimal expansions of rational and irrational numbers.

Giving a decimal expansion of a real number means that given $n \in \mathbb{N}$, we can find $a_0 \in \mathbb{Z}$ and $0 \leq a_1, \cdots, a_n \leq 9$ such that

$$\left|x - \sum_{k=0}^{n} \frac{a_k}{10^k}\right| < \frac{1}{10^n}.$$

In other words if we write

$$x_n = a_0 + \frac{a_1}{10} + \frac{a_2}{10^2} + \cdots + \frac{a_n}{10^n}.$$

then $x_1, x_2, x_3, \cdots, x_n \cdots$ are *approximate* values of x correct to the first, second, third, $\cdots$, nth place of decimal respectively. So when we write a real number by a non-terminating decimal expansion, we mean that we have a scheme of approximation of the real numbers by terminating decimals in such a way that if we stop after the nth place of decimal expansion, then the maximum error committed by us is 10^{-n}.

This brings us to the question of successive approximations of a number. It is obvious that when we have some approximation we ought to have some notion of the error committed. Often we try to reach a number through its approximate values, and the context determines the maximum error admissible. Now if the error admissible is $\varepsilon > 0$, and $x_1, x_2, \cdots, x_n$, is a scheme of successive approximation of a number x, then we should be able to tell at which stage the *desired accuracy* is achieved. In fact, we should find an n such that $|x - x_n| < \varepsilon$. But this could be a chance event. If the error exceeds ε at a later stage, then the scheme cannot be a good scheme of approximation as it is not "stable". Instead it would be desirable that accuracy is achieved at a certain stage and it should not get worse after that stage. This can be realized by demanding that there is a natural number n_0 such that $|x - x_n| < \varepsilon$ for all $n > n_0$. It is clear that n_0 will depend on ε. This leads to the notion of convergence, which is the subject of the next section.

2.5 Sequences and Series

DEFINITION: A sequence is a function $f : \mathbb{N} \to \mathbb{R}$. It is usual to represent the sequence f as $(a_n)_{n=1}^{\infty}$, where $f(n) = a_n$.

DEFINITION: A sequence $(a_n)_{n=1}^{\infty}$ is said to converge to a if for each $\varepsilon > 0$ there is an $n_0 \in \mathbb{N}$ such that

$$|a_n - a| < \varepsilon \quad \text{for all } n > n_0.$$

It can be shown that this number is unique and we write $\lim_{n\to\infty} a_n = a$.

Examples

(a) The sequence $\{1, \frac{1}{2}, \frac{1}{3}, \cdots, \frac{1}{n} \cdots\}$ converges to 0. For $\varepsilon > 0$ let $n_0 = [\frac{1}{\varepsilon}] + 1$. This gives

$$\left|\frac{1}{n} - 0\right| = \frac{1}{n} < \frac{1}{n_0} < \varepsilon \quad \text{for all } n > n_0.$$

(b) The sequence $\{1, \frac{3}{2}, \frac{7}{4}, \frac{15}{8}, \frac{31}{16}, \cdots\}$ converges to 2. The nth term of this sequence is $\frac{2^n - 1}{2^{n-1}} = 2 - \frac{1}{2^{n-1}}$. So $|2 - (2 - \frac{1}{2^{n-1}})| = \frac{1}{2^{n-1}}$. But $2^{n-1} \geq n$ for all $n \geq 1$. Thus, for a given $\varepsilon > 0$, the choice of n_0 given in (a) above will do.

(c) The sequence $(a_n)_{n=1}^{\infty}$, defined by $a_n = n^{\frac{1}{n}}$ converges to 1. Let $n^{1/n} = 1 + \delta_n$

so that for $n > 1$, $\delta_n > 0$. Now $n = (1+\delta_n)^n = 1 + n\delta_n + \frac{n(n-1)}{2}\delta_n^2 + \cdots \geq 1 + \frac{n(n-1)}{2}\delta_n^2$, thus, for $n-1 > 0$, we have $\delta_n \leq \sqrt{\frac{2}{n}}$. For any $\varepsilon > 0$, we can find n_0 in $\mathbb{N}$ such that $n_0\frac{\varepsilon^2}{2} > 1$. Thus for any $n > n_0$ we have $0 \leq \delta_n \leq \sqrt{\frac{2}{n}} \leq \sqrt{\frac{2}{n_0}} < \varepsilon$. This is the same as writing

$$|n^{1/n} - 1| < \varepsilon$$

for $n > n_0$ or equivalently, $\lim_{n\to\infty} n^{1/n} = 1$.

(d) Let $a_n = \frac{2^n}{n!}$. The sequence $(a_n)_{n=1}^{\infty}$ converges to 0. Note that

$$a_n = \frac{2}{1}\cdot\frac{2}{2}\cdot\frac{2}{3}\cdot\cdots\cdot\frac{2}{n} < \frac{4}{n} \text{ for } n > 3.$$

For $\varepsilon > 0$, choose n_1 such that $n_1\varepsilon > 4$. Now let $n_0 = \max\{3, n\}$, so that $|a_n| < \varepsilon$ for all $n > n_0$.

(e) For $a_n = \frac{n!}{n^n}$ the sequence $(a_n)_{n=1}^{\infty}$ converges to 0. **(Exercise!)**

In all the above examples we somehow guessed in advance what a sequence converges to. But suppose we are not able to do that and one asks whether it is possible to decide if the sequence converges to some real number. This can be helped by the following analogy: Suppose there are many people coming to Delhi to attend a conference. They might be taking different routes. But as they come closer and closer to Delhi the distance between the participants is getting smaller. This can be paraphrased in mathematical language as:

THEOREM: *If $(a_n)_{n=1}^{\infty}$ is a convergent sequence, then for every $\varepsilon > 0$ we can find an n_0 such that*

$$|a_n - a_m| < \varepsilon \quad \textit{for all} \quad m, n > n_0.$$

PROOF: Suppose $(a_n)_{n=1}^{\infty}$ converges to a. Then for every $\varepsilon > 0$ we can find an n_0 such that

$$|a_n - a| < \frac{\varepsilon}{2} \quad \text{for all} \quad n > n_0.$$

So for $m, n > n_0$, we have

$$|a_m - a_n| = |a_n - a + a - a_m| \leq |a_n - a| + |a - a_m| < \varepsilon. \qquad \blacksquare$$

The proof is rather simple. This very useful idea was conceived by Cauchy and the above theorem is called the Cauchy criterion for convergence. What we have proved above tells us that the criterion is a necessary condition for convergence. Is it also sufficient? That is, given a sequence $(a_n)_{n=1}^{\infty}$ which satisfies the Cauchy criterion, can we assert that there is a real number to which it converges? The answer is yes!

We call a sequence $(a_n)_{n=1}^{\infty}$ monotonically non-decreasing if $a_{n+1} \geq a_n$ for all n.

THEOREM: *A monotonically non-decreasing sequence which is bounded above converges.*

PROOF: Suppose $(a_n)_{n=1}^{\infty}$ is monotonic and non-decreasing. We have $a_1 \leq a_2 \leq \cdots \leq a_n \leq a_{n+1} \leq \cdots$. Since the sequence is bounded above, $\{a_k : k = 1, \cdots\}$ has a least upper bound. Let it be a. By definition, $a_n \leq a$ for all n, but for $\varepsilon > 0$ there is at least one n_0 such that $a_{n_0} + \varepsilon > a$. Therefore for $n > n_0$, $a_n + \varepsilon \geq a_{n_0} + \varepsilon > a \geq a_n$. This gives us $|a_n - a| < \varepsilon$ for all $n \geq n_0$. ■

We can similarly prove that: A monotonically non-increasing sequence which is bounded below is convergent.

Suppose we did not have the condition of boundedness below or above for a monotonically non-increasing or non-decreasing sequence respectively, then what would happen? If a sequence is monotonically non-decreasing and is not bounded above, then given any real number $M > 0$ there exists at least one n_0 such that $a_{n_0} > M$, and hence $a_n > M$ for all $n > n_0$. In such a case we say that $(a_n)_{n=1}^{\infty}$ *diverges* to ∞. We write $\lim\limits_{n\to\infty} a_n = \infty$. More generally (i.e., even when the sequence is not monotone), the same criterion as above allows us to say that $\{a_n\}_{n=1}^{\infty}$ diverges to ∞ and we write $\lim\limits_{n\to\infty} a_n = \infty$. We can similarly define divergence to $-\infty$.

We make a digression here: In case a set is bounded above, then we have the concept of least upper bound. For any set A of real numbers, we define *supremum* of A as

$$\begin{aligned}\sup\{x : x \in A\} &= \sup A \\ &= \begin{cases} \text{least upper bound of } A & \text{if } A \text{ is bounded above,} \\ \infty & \text{if } A \text{ is not bounded above.} \end{cases}\end{aligned}$$

Similarly we define *infimum* of a set A as

$$\begin{aligned}\inf\{x : x \in A\} &= \inf A \\ &= \begin{cases} \text{greatest lower bound of } A, & \text{if } A \text{ is bounded below,} \\ -\infty & \text{if } A \text{ is not bounded below.} \end{cases}\end{aligned}$$

This would help us to look for other criteria of convergence. For any bounded sequence $(a_n)_{n=1}^{\infty}$ of real numbers let

$$b_n = \sup\{a_n, a_{n+1}, a_{n+2}, \cdots\} = \sup\{a_k : k \geq n\}.$$

It is clear that $(b_n)_{n=1}^{\infty}$ is now a non-increasing sequence. So it converges. We set

$$\begin{aligned}\lim_{n\to\infty} b_n &= \text{limit superior of the sequence } (a_n)_{n=1}^{\infty}\\ &= \limsup a_n = \overline{\lim}_n a_n\end{aligned}$$

Similarly if we write

$$c_n = \inf\{a_n, a_{n+1}, \cdots\} = \inf\{a_k : k \geq n\},$$

then $(a_n)_{n=1}^{\infty}$ is a monotonically non-decreasing sequence. So we write

$$\begin{aligned}\lim_{n\to\infty} c_n &= \text{limit inferior of the sequence } (a_n)_{n=1}^{\infty}\\ &= \liminf a_n = \underline{\lim} a_n.\end{aligned}$$

We may not know a sequence to be convergent or divergent, yet we can calculate its limit superior and limit inferior.

We have in fact the following result:

THEOREM: *For a sequence $(a_n)_{n=1}^{\infty}$ of real numbers,*

$$\underline{\lim}\, a_n \leq \overline{\lim}\, a_n.$$

Further, if $\ell = \underline{\lim}\, a_n$ and $L = \overline{\lim}\, a_n$ are finite, then $\ell = L$ if and only if the sequence is convergent.

PROOF: It is easy to see that $\ell \leq L$. Now, suppose ℓ, L are finite. If $\ell = L$, then for all $\varepsilon > 0$ there exist n_1, n_2 such that

$$\ell - \varepsilon < \sup_{k\geq n} a_k < \ell + \varepsilon \text{ for all } n > n_1 \text{ and } \ell - \varepsilon < \inf_{k\geq n} a_k < \ell + \varepsilon \text{ for all } n > n_2.$$

Thus with $n_0 = \max\{n_1, n_2\}$ we have

$$\ell - \varepsilon < a_n < \ell + \varepsilon \quad \text{for all} \quad n > n_o.$$

This proves that equality holds only when it is convergent. Conversely, suppose

a_n converges to a. For every $\varepsilon > 0$, we have n_0 such that $a - \varepsilon < a_n < a + \varepsilon$ for all $n > n_0$. Therefore, $\sup_{k \geq n} a_k \leq a + \varepsilon$ and $a - \varepsilon \leq \inf a_n$ for all $n > n_0$. Hence $a - \varepsilon \leq \ell < L \leq a + \varepsilon$. Since this is true for every $\varepsilon > 0$ we have $L = \ell = a$. ■

DEFINITION: We call a sequence $(a_n)_{n=1}^{\infty}$ a Cauchy sequence if for all $\varepsilon > 0$ there exists an n_0 such that

$$|a_m - a_n| < \varepsilon \quad \text{for all } m, n > n_0.$$

THEOREM: *Every Cauchy sequence is a bounded sequence and is convergent.*

PROOF: By definition, for all $\varepsilon > 0$ there is an n_0 such that

$$|a_m - a_n| < \varepsilon, \quad \text{for all } m, n > n_0.$$

So, in particular, $|a_{n_0+1} - a_n| < \varepsilon$, for all $n > n_0$, i.e.,

$$a_{n_0+1} - \varepsilon < a_n < a_{n_0+1} + \varepsilon, \quad \text{for all } n > n_0.$$

Let $M = \max\{a_1, \cdots, a_{n_0}, a_{n_0+1} + \varepsilon\}$ and $m = \min\{a_1, \cdots, a_{n_0}, a_{n_0+1} - \varepsilon\}$. It is clear that

$$m \leq a_n \leq M, \quad \text{for all } n \geq 1.$$

We now prove that such a sequence is convergent. Let $\overline{\lim}\, a_n = L$ and $\underline{\lim}\, a_n = \ell$. Since any Cauchy sequence is bounded,

$$-\infty < \ell \leq L < \infty.$$

But since $(a_n)_{n=1}^{\infty}$ is Cauchy, for every $\varepsilon > 0$ there is an $n_0 = n_0(\varepsilon)$ such that

$$a_{n_0+1} - \varepsilon < a_n < a_{n_0+1} + \varepsilon, \quad \text{for all } n > n_0,$$

which implies that $a_{n_0+1} - \varepsilon \leq \underline{\lim}\, a_n = \ell \leq \overline{\lim}\, a_n = L \leq a_{n_0+1} + \varepsilon$. Thus $L - \ell \leq 2\varepsilon$, for all $\varepsilon > 0$. This is possible only if $L = \ell$. ■

Thus we have established that the Cauchy criterion is both a necessary and sufficient criterion of convergence of a sequence. We state a few more results without proofs (**Exercise**).

THEOREM: *For sequences* $(a_n)_{n=1}^{\infty}$ *and* $(b_n)_{n=1}^{\infty}$,

(i) *If* $l \leq a_n \leq b_n$ *and* $\lim\limits_{n\to\infty} b_n = l$, *then* $(a_n)_{n=1}^{\infty}$ *too is convergent and* $\lim\limits_{n\to\infty} a_n = \ell$.

(ii) *If* $a_n \leq b_n$, *then* $\overline{\lim}\ a_n \leq \overline{\lim}\ b_n$, $\underline{\lim}\ a_n \leq \underline{\lim}\ b_n$.

(iii) $\underline{\lim}\ (a_n + b_n) \geq \underline{\lim}\ a_n + \underline{\lim}\ b_n$.

(iv) $\overline{\lim}\ (a_n + b_n) \leq \overline{\lim}\ a_n + \overline{\lim}\ b_n$.

(v) *If* $(a_n)_{n=1}^{\infty}$ *and* $(b_n)_{n=1}^{\infty}$ *are both convergent, then* $(a_n + b_n)_{n=1}^{\infty}$, $(a_n - b_n)_{n=1}^{\infty}$ *and* $(a_n b_n)_{n=1}^{\infty}$ *are convergent and we have* $\lim\ (a_n \pm b_n) = \lim a_n \pm \lim b_n$ *and* $\lim a_n b_n = \lim a_n \lim b_n$.

(vi) *If* $(a_n)_{n=1}^{\infty}, (b_n)_{n=1}^{\infty}$ *are convergent and* $\lim\limits_{n\to\infty} b_n = \ell \neq 0$, *then* $\left(\frac{a_n}{b_n}\right)_{n=1}^{\infty}$ *is convergent and* $\lim\limits_{n\to\infty} \dfrac{a_n}{b_n} = \dfrac{\lim a_n}{\lim b_n}$.

Some Useful Limits

Consider the sequence $\left(\left(1+\frac{1}{n}\right)^n\right)_{n=1}^{\infty}$. Does it converge? Binomial expansion gives

$$\begin{aligned}
a_n &= \left(1+\frac{1}{n}\right)^n = 1 + n\cdot\frac{1}{n} + \frac{n(n-1)}{2!}\frac{1}{n^2} \\
&\quad + \frac{n(n-1)(n-2)}{3!}\cdot\frac{1}{n^3} + \cdots + \frac{1}{n^n} \\
&= 1 + 1 + \frac{\left(1-\frac{1}{n}\right)}{2!} + \frac{\left(1-\frac{1}{n}\right)\left(1-\frac{2}{n}\right)}{3!} + \cdots + \\
&\quad \frac{\left(1-\frac{1}{n}\right)\cdots\left(1-\frac{n-2}{n}\right)}{(n-1)!} + \frac{\left(1-\frac{1}{n}\right)\cdots\left(1-\frac{n-1}{n}\right)}{n!}
\end{aligned}$$

$$\begin{aligned}
a_{n+1} &= 1 + 1 + \frac{\left(1-\frac{1}{n+1}\right)}{2!} + \frac{\left(1-\frac{1}{n+1}\right)\left(1-\frac{2}{n+1}\right)}{3!} + \cdots + \\
&\quad \frac{\left(1-\frac{1}{n+1}\right)\cdots\left(1-\frac{n-1}{n+1}\right)}{n!} + \frac{\left(1-\frac{1}{n+1}\right)\cdots\left(1-\frac{n}{n+1}\right)}{(n+1)!}
\end{aligned}$$

So we have $a_n \leq a_{n+1}$.

But

$$\begin{aligned} a_n &< 1+1+\frac{1}{2!}+\frac{1}{3!}+\cdots+\frac{1}{n!} \\ &< 1+1+\frac{1}{2}+\frac{1}{4}+\cdots+\frac{1}{2^{n-1}} \\ &= 1+1+\left(1-\frac{1}{2^{n-1}}\right) \\ &= 3-\frac{1}{2^{n-1}} < 3. \end{aligned}$$

Thus we have $(a_n)_{n=1}^{\infty}$, a monotonically increasing sequence which is bounded above. Hence, it must converge to a limit. This limit is usually denoted by e, i.e.,

$$\lim_{n\to\infty}\left(1+\frac{1}{n}\right)^n = e.$$

It is easily seen that e is a number between 2 and 3. It is an irrational number (to be shown later) and plays a very important role in mathematics. Often it is convenient to take e as the base of logarithm. We shall come back to it a little later.

We had earlier seen the need for extending $\mathbb{Q}$ to $\mathbb{R}$. One of the reasons was that we could define ${}^n\sqrt{x}$ when $x \geq 0$ and n is a positive integer. We write $\sqrt[n]{x} = x^{\frac{1}{n}}$, so we define $x^{p/q} = \sqrt[q]{x^p}$ because that would facilitate us to have the law of indices

$$x^r x^s = x^{r+s}$$

when r and s are rational numbers. But can one define x^a when a is real but not necessarily a rational number? For example, how can one define $2^{\sqrt{2}}$?

We proceed thus. For a fixed positive real number x we can define $x^{\frac{p}{q}} = \sqrt[q]{x^p}$ if p, q are positive integers; and if $\frac{p}{q}$ is negative, then $-\frac{p}{q}$ is positive rational and we define $\frac{1}{x^{(-\frac{p}{q})}}$ as $x^{\frac{p}{q}}$.

For any real number $a > 0$ and $x > 1$, let us write

$$E = \{x^r : r \in \mathbb{Q} \text{ and } 0 < r < a\}.$$

Since $x^r < x^s$ for $r < s$, $r, s \in \mathbb{Q}$, it is obvious that E is bounded above. So E has a least upper bound. We call this least upper bound x^a. It is easy to see that $x^a > 1$. For $a = 0$, x^0 is defined as 1 for all $x > 0$. If $a < 0$, x^a for $x > 1$ is defined as shown below. Letting $b = -a$, we define

$$x^a = \frac{1}{x^b}.$$

When $x = 1$, 1^a is defined as 1 for all a. Finally, for $0 < x < 1$, x^a is defined as

$$x^a = \left(\frac{1}{x}\right)^{-a}$$

We can easily verify that (**Exercise**):

(i) $a^x a^y = a^{x+y}$,

(ii) if $a > 1$, then $a^x > 1$ for $x > 0$ and if $a < 1$, then $a^x < 1$ for $x > 0$,

(iii) $a^x < a^y$ for $x < y$ if $a > 1$ and $a^x > a^y$ for $x < y$ if $a < 1$,

(iv) $a^x \div a^y = a^{x-y}$.

Series

Now consider $a_n = 1 - \frac{1}{3} + \frac{1}{5} - \frac{1}{7} + \cdots + \frac{1}{4n-3} - \frac{1}{4n-1}$. Then $(a_n)_{n=1}^{\infty}$ is easily seen to be monotonically increasing, because

$$a_{n+1} - a_n = \frac{1}{4n+1} - \frac{1}{4n+3} \geq 0.$$

Again, each $a_n \geq 0$ as $(1 - \frac{1}{3}) \geq 0$, $(\frac{1}{5} - \frac{1}{7}) \geq 0, \cdots$ and

$$a_n = 1 - \left[\left(\frac{1}{3} - \frac{1}{5}\right) + \left(\frac{1}{7} - \frac{1}{9}\right) + \cdots + \left(\frac{1}{4n-5} - \frac{1}{4n-3}\right) + \frac{1}{4n-1}\right],$$

so $a_n < 1$. Thus we have a monotonically increasing sequence bounded above. It must converge to a limit which is known to be $\frac{\pi}{4}$, i.e., $\lim_{n\to\infty} a_n = \frac{\pi}{4}$. Thus $a_1, a_2, a_3, \cdots, a_n$ may be looked upon as a scheme of approximation for $\frac{\pi}{4}$. The formula $\frac{\pi}{4} = 1 - \frac{1}{3} + \frac{1}{5} - \frac{1}{7} + \frac{1}{9} - \frac{1}{11} + \cdots$ is called Madhava-Gregory series. Recall the definition of convergence. For $\varepsilon > 0$ we must be able to find n_0 such that

$$\left|\frac{\pi}{4} - a_n\right| < \varepsilon \quad \text{for all} \quad n > n_0.$$

In fact, it can be shown that $\left|\frac{\pi}{4} - a_n\right| < \frac{1}{2n+1}$. This should give us a method of computing π to any desired accuracy. If we write a program to compute a_n to approximate π accurate within ε, then n_0 is the minimum number of steps for the computation to work.

In the last example, we came across an infinite sum of real numbers, called series. We know how to add finitely many real numbers. But what does it mean to add infinitely many real numbers?

An expression like $a_1 + a_2 + \cdots + a_n + \cdots$ is called an infinite series and is often abbreviated by $\sum_{n=1}^{\infty} a_n$, where $a_1, a_2, \cdots, a_n, \cdots$ are real numbers.

Let us write $s_1 = a_1$, $s_2 = a_1 + a_2$, $s_3 = a_1 + a_2 + a_3 \cdots$, $s_n = a_1 + \cdots + a_n$, called the partial sums of the series $\sum_{n=1}^{\infty} a_n$. Now the partial sums constitute a sequence $(s_n)_{n=1}^{\infty}$ of real numbers. If this sequence is convergent, then we say that the *series* $\sum_{n=1}^{\infty} a_n$ *is convergent* and the real number to which the sequence of partial sums converges is called the *sum* of the series. If $(s_n)_{n=1}^{\infty}$ is divergent, then we say that the corresponding series is divergent.

Examples

(a) Consider the series $\sum_{n=1}^{\infty} \frac{1}{2^n}$. In this case $s_1 = \frac{1}{2}$, $s_2 = \frac{3}{4}$, $s_3 = \frac{7}{8}$, $s_4 = \frac{15}{16}, \cdots, s_n = \frac{2^n - 1}{2^n} = 1 - \frac{1}{2^n}$ and it is easy to see that the sequence $(s_n)_{n=1}^{\infty}$ converges to 1.

(b) Consider the series $\frac{1}{1.2} + \frac{1}{2.3} + \frac{1}{3.4} + \frac{1}{4.5} + \cdots + \cdots$. Here $a_1 = \frac{1}{1.2}$, $a_2 = \frac{1}{2.3}, \cdots$.
But $a_1 = 1 - \frac{1}{2}$, $a_2 = \frac{1}{2} - \frac{1}{3}$, $a_3 = \frac{1}{3} - \frac{1}{4}, \cdots, a_n = \frac{1}{n} - \frac{1}{n+1}$. Therefore $s_n = a_1 + a_2 + \cdots + a_n = 1 - \frac{1}{n+1}$ and hence $(s_n)_{n=1}^{\infty}$ converges to 1.

(c) Consider the series $\sum_{n=1}^{\infty} n$. Here $s_n = 1 + 2 + \cdots + n = \frac{n(n+1)}{2}$. It is easy to see that $(s_n)_{n=1}^{\infty}$ is a divergent sequence diverging to ∞.

These are examples of series where the nth partial sum has a nice formula. But often we are not so fortunate and it is quite impractical to get the sequence of partial sums and then decide if it is convergent or not. However we may state:

THEOREM: *A necessary and sufficient condition that the series $\sum_{n=1}^{\infty} a_n$ is convergent is that for each $\varepsilon > 0$ there is an n_0 such that*

$$\left| \sum_{k=m+1}^{n} a_k \right| < \varepsilon \text{ for all } \quad n > m > n_0.$$

It is clear that this is nothing but the Cauchy criterion of convergence applied to the sequence of partial sums of the series $\sum_{n=1}^{\infty} a_n$. In particular, if $\sum_{n=1}^{\infty} a_n$ is convergent, then $\lim\limits_{n \to \infty} a_n = 0$.

Warning: The converse need not be true as the following example shows.

Examples

(a) The series $\sum_{n=1}^{\infty} \frac{1}{n}$ diverges. This is not convergent since, $\sum_{k=m+1}^{2m} \frac{1}{k} = \frac{1}{m+1} + \frac{1}{m+2} + \cdots + \frac{1}{2m} > \frac{1}{2}$ for any m. So there is no way in which we can find an n_0 beyond which $\sum_{k=m+1}^{2m} \frac{1}{k}$ would be less than even $\frac{1}{2}$, let alone arbitrarily small ε.

(b) Consider the series $\sum_{n=1}^{\infty} \frac{1}{n^2}$. It is easy to see that

$$\left|\sum_{k=m+1}^{n} \frac{1}{k^2}\right| < \left|\sum_{k=m+1}^{n} \frac{1}{(k-1)k}\right| = \left|\sum_{k=m}^{n-1} \frac{1}{k(k+1)}\right| = \left|\frac{1}{m} - \frac{1}{n}\right| \leq \frac{1}{m} + \frac{1}{n}.$$

Thus for $m, n > n_0 > 2/\varepsilon$, we have that

$$\left|\sum_{k=m+1}^{n} \frac{1}{k^2}\right| < \varepsilon.$$

(c) $\sum_{n=0}^{\infty} \frac{1}{n!}$ is convergent since for $m \geq 2$,

$$\sum_{k=m+1}^{n} \frac{1}{k!} = \frac{1}{(m+1)!} + \frac{1}{(m+2)!} + \cdots + \frac{1}{n!} < \frac{1}{2^m} + \frac{1}{2^{m+1}} + \cdots + \frac{1}{2^{n-1}}$$

$$< \frac{1}{2^m}\left(\frac{1 - \frac{1}{2^{n-m}}}{1 - \frac{1}{2}}\right) < \frac{1}{2^{m-1}},$$

and since we can find for $\varepsilon > 0$ an n_0 such that $\frac{1}{2^{m-1}} < \varepsilon$ for $m > n_0$.

The Cauchy criterion of convergence of a series also leads to the following:

THEOREM: *Convergence of a series is not affected by changing finitely many terms of the series.*

The proof of this is obvious because we are interested only in the tail end $\sum_{k=m+1}^{n} a_k$ of the series which is unaffected by changing finitely many terms when m is large enough. Despite the elegance of the Cauchy criterion of convergence, its direct use is fraught with difficulties. The above proposition, however, tells us that if we know that two series differ only in finitely many terms, then the convergence of one implies that of the other.

Exercise: Prove that the convergence of a series is not affected by multiplying each of its terms by a fixed non-zero real number.

We give below a few sufficient conditions of convergence.

COMPARISON TEST: *If $a_n \geq 0$, $b_n \geq 0$ for all n and $a_n \leq b_n$ for all but finitely many n, then convergence of $\sum b_n$ implies convergence of $\sum a_n$.* **(Exercise)**

We have a result complementary to the above. Suppose that $a_n, b_n \geq 0$ and that $a_n \leq b_n$ for all except finitely many n. If $\sum a_n$ is divergent, then $\sum b_n$ is divergent.

Examples

(a) $\sum \frac{1}{n^2}$ is convergent since $\frac{1}{n^2} < \frac{1}{(n-1)n}$ for $n \geq 2$, and $\sum_{n=2}^{\infty} \frac{1}{n(n-1)}$ is convergent.

(b) $\sum(\sqrt{n^2+1} - n)$ is divergent, since $\sqrt{n^2+1} - n = \frac{1}{\sqrt{n^2+1}+n} > \frac{1}{2(n+1)}$ and $\sum \frac{1}{n+1}$ is divergent.

(c) Since $\frac{1}{n!} < \frac{1}{2^n} (n \geq 2)$ and since the series $\sum \frac{1}{2^n}$ is convergent, the series $1 + \frac{1}{1!} + \frac{1}{2!} + \cdots + \frac{1}{n!} + \cdots$ is convergent.

In fact, we can say what it converges to. We have seen that $\left(1 + \frac{1}{n}\right)^n$ converges to a number which we called e. Now, by the binomial theorem,

$$\begin{aligned}\left(1 + \tfrac{1}{n}\right)^n &= 1 + \tfrac{1}{1!} + \frac{(1-\frac{1}{n})}{2!} + \frac{(1-\frac{1}{n})(1-\frac{2}{n})}{3!} + \cdots + \frac{(1-\frac{1}{n})(1-\frac{2}{n})\cdots(1-\frac{n-1}{n})}{n!} \\ &< 1 + \tfrac{1}{1!} + \tfrac{1}{2!} + \cdots + \tfrac{1}{n!}.\end{aligned}$$

Thus $e \leq \sum_{n=1}^{\infty} \frac{1}{n!}$. But

$$\left(1 + \frac{1}{n}\right)^n \geq 1 + \frac{1}{1!} + \frac{(1-\frac{1}{n})}{2!} + \cdots + \frac{(1-\frac{1}{n})\cdots(1-\frac{m-1}{n})}{m!} \text{ for every fixed } m < n.$$

Again taking limit $n \to \infty$, we have

$$e \geq 1 + \frac{1}{1!} + \frac{1}{2!} + \cdots + \frac{1}{m!} \text{ for every } m.$$

This gives us $e = 1 + \frac{1}{1!} + \cdots + \frac{1}{n!} + \cdots$.

From the comparison test we have the following:

COROLLARY: *Suppose $a_n, b_n > 0$ and suppose furthermore that $\frac{a_{n+1}}{a_n} \leq \frac{b_{n+1}}{b_n}$ for all but finitely many n. Then convergence of $\sum b_n$ implies convergence of $\sum a_n$. Conversely, divergence of $\sum a_n$ implies divergence of $\sum b_n$.*

PROOF: Suppose $\frac{a_{n+1}}{a_n} \le \frac{b_{n+1}}{b_n}$ for $n \ge n_0$. Then we have

$$\begin{aligned} a_n &= \frac{a_n}{a_{n-1}}\frac{a_{n-1}}{a_{n-2}}\cdots\frac{a_{n_0+1}}{a_{n_0}}a_{n_0} \\ &\le \frac{b_n}{b_{n-1}}\frac{b_{n-1}}{b_{n-2}}\cdots\frac{b_{n_0+1}}{b_{n_0}}a_{n_0} \\ &= \frac{a_{n_0}}{b_{n_0}}b_n. \end{aligned}$$

So by the comparison test we have the result. ■

Let $\sum a_n$ be a series of positive terms.

(a) Let $0 < p < 1$ and suppose there is an $n_0 \in \mathbb{N}$ such that $\frac{a_{n+1}}{a_n} < p$ for all $n > n_0$, then $\sum a_n$ is convergent.

(b) Let $p \ge 1$ and suppose that there is an n_0 such that $\frac{a_{n+1}}{a_n} > p$ for all $n > n_0$, then $\sum a_n$ is divergent.

The proof follows from the fact that $\sum_{n=0}^{\infty} p^n$ converges if $p < 1$ and diverges if $p \ge 1$.

These two can be combined into:

D'ALEMBERT'S RATIO TEST: *Suppose for a series of positive terms $\sum a_n$, $\lim_{n\to\infty} \frac{a_{n+1}}{a_n} = l$. If $l < 1$ the series converges, and if $l > 1$ the series diverges. If $l = 1$ the test fails.*

PROOF: (a) if $l < 1$, let $p = \frac{l+1}{2}$ and $\varepsilon = \frac{1-l}{2}$. Now since $\lim_{n\to\infty} \frac{a_{n+1}}{a_n} = l$, we can get an n_o such that $l - \varepsilon < \frac{a_{n+1}}{a_n} < l + \varepsilon = p < 1$ for $n > n_0$.

So we can apply the above test.

(b) Similarly, for $l > 1$, let $p = \frac{l+1}{2}$ and $\varepsilon = \frac{l-1}{2}$. Since $\lim \frac{a_{n+1}}{a_n} = l$, we have an n_0 such that $1 - \varepsilon < \frac{a_{n+1}}{a_n}$ for all $n > n_0$ and $1 < p < \frac{a_{n+1}}{a_n}$ for all $n > n_0$. So $\sum a_n$ is divergent.

In case $l = 1$ we cannot apply either of the tricks (a) or (b). Indeed for $a_n = \frac{1}{n^2}$, $\lim \frac{a_{n+1}}{a_n} = 1$ yet $\sum a_n$ is convergent, whereas for $a_n = \frac{1}{n}$, $\lim \frac{a_{n+1}}{a_n} = 1$ yet $\sum a_n$ is divergent. This shows that the test cannot be applied if $l = 1$. ■

CAUCHY'S ROOT TEST: *Let $a_n \ge 0$.*
(a) Suppose there is a positive number $p < 1$ such that $(a_n)^{\frac{1}{n}} \le p$ for all but finitely many terms. Then $\sum a_n$ is convergent.

(b) Suppose there is a $p \geq 1$ and $(a_n)^{\frac{1}{n}} \geq p$ for all but finitely many n. Then $\sum a_n$ is divergent.

This follows by comparing a_n with p^n. The above two combine to give the following test:

Exercises

Suppose $a_n \geq 0$ and $\lim\limits_{n\to\infty} (a_n)^{\frac{1}{n}} = l$.

(a) if $l < 1$, then the series $\sum a_n$ is convergent,

(b) if $l > 1$, then the series $\sum a_n$ is divergent,

(c) if $l = 1$ the test fails.

Often $\lim\limits_{n\to\infty} (a_n)^{\frac{1}{n}}$ may not exist. In such a case we have for $a_n \geq 0$

(a) if $\overline{\lim}(a_n)^{\frac{1}{n}} = l < 1$, then $\sum a_n$ is convergent.

(b) if $\underline{\lim}(a_n)^{\frac{1}{n}} = l > 1$, then $\sum a_n$ is divergent.

Examples

(a) Consider $\sum_{n=1}^{\infty} \frac{x^n}{n}$, $x \geq 0$. In this case $a_n = \frac{x^n}{n}$, $\lim \frac{a_{n+1}}{a_n} = \lim x\frac{n}{n+1} = \lim \frac{x}{1+\frac{1}{n}} = x$. So the series converges if $x < 1$ and diverges if $x > 1$. For $x = 1$, the series is $\sum \frac{1}{n}$, which diverges as we have seen earlier.

(b) Consider $\sum_{n=1}^{\infty} \left(1 + \frac{1}{n}\right)^{-n^2}$. Since $\lim(\, a_n)^{\frac{1}{n}} = \lim\limits_{n\to\infty} \dfrac{1}{(1+\frac{1}{n})^n} = \dfrac{1}{e} < 1$, $\sum_{n=1}^{\infty}(1+\frac{1}{n})^{-n^2}$ is convergent.

(c) Consider the series $\frac{1}{2} + \frac{1}{3} + \frac{1}{2^2} + \frac{1}{3^2} + \frac{1}{2^3} + \frac{1}{3^3} + \cdots$. Obviously, neither $\lim \frac{a_{n+1}}{a_n}$ nor $\lim a_n^{1/n}$ exist. But $\overline{\lim}\ a_n^{1/n} = \frac{1}{\sqrt{2}} < 1$. Hence the series is convergent **(Exercise!)**.

Hitherto we have been dealing only with a series of positive terms. What about a series whose terms are of both positive and negative signs? One way to deal with such a series is to consider the series of absolute values of the terms of the original series. We have:

DEFINITION: The series $\sum a_n$ is called an *absolutely convergent* series if $\sum |a_n|$ is convergent.

How does this definition help us? What bearing can it have on the given series? We have indeed:

THEOREM: *An absolutely convergent series is convergent.*

PROOF: Suppose $\sum |a_n|$ is convergent. We have to show that $\sum a_n$ is convergent. Let $t_n = \sum_{k=1}^{n} |a_k|$ and $s_n = \sum_{k=1}^{n} a_k$.

It is easy to see that $|s_m - s_n| \leq |t_m - t_n|$ for all m, n. Since $\sum_{n=1}^{\infty} |a_n|$ is convergent, we have for all $\varepsilon > 0$, there exists an n_0 such that $|t_m - t_n| < \varepsilon$ for $m, n > n_0$.

So $|s_m - s_n| \leq |t_m - t_n| < \varepsilon$ for all $m, n > n_0$. Hence by the Cauchy criterion of convergence, $\sum a_n$ is convergent. ∎

Example: $1 - \frac{1}{2^2} + \frac{1}{3^2} - \frac{1}{4^2} + \cdots$ is convergent because it is absolutely convergent.

We have shown that an absolutely convergent series is convergent. But that does not mean that every convergent series is absolutely convergent.

ALTERNATING SERIES TEST: *Suppose* $a_1 \geq a_2 \geq a_3 \geq \cdots \geq a_n \geq a_{n+1} \geq \cdots \geq 0$ *and* $\lim_{n \to \infty} a_n = 0$. *Then the series* $a_1 - a_2 + a_3 - a_4 + a_5 + \cdots$ *is convergent.*

PROOF: Let $t_n = s_{2n} = a_1 - a_2 + a_3 - a_4 + \cdots + a_{2n-1} - a_{2n} \geq 0$. It is obvious that $t_{n+1} \geq t_n$. Thus $(t_n)_{n=1}^{\infty}$ is a monotonically increasing sequence. But

$$t_n = a_1 - ((a_2 - a_3) + (a_4 - a_5) + \cdots + (a_{2n-2} - a_{2n-1}) + a_{2n}).$$

Since each of $a_2 - a_3$, $a_4 - a_5, \cdots, a_{2n-2} - a_{2n-1}$ and a_{2n} is non-negative, it follows that $t_n \leq a_1$.

Thus $(t_n)_{n=1}^{\infty}$ is a convergent sequence. Let $\lim t_n = \lim s_{2n} = s$. Now $s_{2n+1} = s_{2n} + a_{2n+1}$, so that $\lim s_{2n+1} = \lim s_{2n} + \lim a_{2n+1} = s$. Thus, $(s_n)_{n=1}^{\infty}$ is convergent. ∎

The alternating series test applies to the series $1 - \frac{1}{2} + \frac{1}{3} - \frac{1}{4} + \frac{1}{5} \cdots$ and hence it is convergent. But it is not absolutely convergent as $1 + \frac{1}{2} + \frac{1}{3} + \frac{1}{4} + \cdots + \frac{1}{n} + \cdots$ is divergent. A series which is convergent but not absolutely convergent is called a *conditionally convergent* series.

Product of Series

Let $\sum_{n=1}^{\infty} a_n$ and $\sum_{n=1}^{\infty} b_n$ be absolutely convergent series and let $c_n = a_n b_1 + a_{n-1}b_2 + \cdots + a_1 b_n = \sum_{k=0}^{n-1} a_{n-k} b_{k+1}$. Then $\sum c_n$ is absolutely convergent, and $\sum c_n = (\sum a_n)(\sum b_n)$.

But if neither Σa_n nor Σb_n is absolutely convergent, then Σc_n might not converge. Consider for example

$$a_n = \frac{(-1)^{n-1}}{\sqrt{n}}, \quad b_n = \frac{(-1)^n}{\sqrt{n}}.$$

Σa_n and Σb_n are convergent by the alternating series test but not absolutely convergent. Compute $c_n = \sum_{k=0}^{n} a_{n-k} b_{k+1}$ and verify that c_n *does not* converge to 0 as $n \to \infty$.

Historical Notes

It was known to early civilization that the circumference of a circle bears a constant ratio to its diameter (irrespective of the size of the circle). This constant is denoted in modern times by the Greek letter π (Euler was the first one to have used this notation). This has been an enigma to people since antiquity. The old testament suggested its value as 3, whereas the ancient Egyptians assigned it the value $3\frac{1}{6}$. Archimedes for the first time suggested a geometric method of determining the approximate value of π to any degree of accuracy; his calculation led to the inequality $3\frac{10}{71} < \pi < 3\frac{10}{70}$. The later approximate value has been so popular that π is *mistakenly* taken to be exactly equal to $\frac{22}{7}$. Aryabhatta (476–550 AD) took its value to be 3.1416 but never suggested how he got it. Bhaskara (1114–1185 AD) was the first one who hinted at the irrationality of π. Madhava (1340–1425 AD) actually gave the series expansion of arctangent in terms of its arguments which leads to the value of π. Nilakantha (1443–1543 AD) confirmed that π must be irrational. Gregory rediscovered the series of Madhava. The best approximation scheme for numerical computation of the value of π was by the legendary self-taught Indian mathematician Srinivasa Ramanujan (1887–1920).

Suggested Reading

1. *Calculus with Applications and Computing* by P. Lax, S. Burstein and A. Lax, Narosa Publishing House, New Delhi 1979. (read Chapter 1).

2. *Calculus*, Vol. I by T.M. Apostol, Wiley Eastern, New Delhi 1980 (good Introduction).

Chapter 3

Continuity, Differentiability and Integration

"I could be bounded in a nutshell and count myself a king of infinite space."

William Shakespeare, (Hamlet, Act II, Scene 2)

3.1 Limits and Continuity

We have seen many examples of functions in Chapter 1. Let us also consider the example of the range r of a gun pointed at an angle θ to the horizon. Gun, shell, and other conditions remaining the same, we know that for every θ we have a definite range $r(\theta)$, giving rise to a function r. Suppose we know that the target is at a distance d from the gun and that $r(\theta_0) = d$ for some θ_0. Then we point our gun at an angle θ_0 to the horizon to hit the target. A little deviation in θ is likely to cause an error in our hit. But we also know that if our hit is within a certain distance from the target, then our objective is achieved (depending on the shell). Now, if we want the hit to be within a distance $\varepsilon > 0$ from the target, can we adjust the deviation δ in θ around θ_0 accordingly? Another way of putting the same question is to ask if for a small change in angle θ, can one get a small change in the range $r(\theta)$? If yes, then we are allowed a little play in aiming our gun. If not, a small play might result in a wide miss. What we are asking is: for every $\varepsilon > 0$, can we find a $\delta > 0$ such that $|r(\theta) - r(\theta_0)| < \varepsilon$ whenever $|\theta - \theta_0| < \delta$? We shall come across similar questions in other situations too. For example, we have described earlier how to obtain the value of π through the series called Madhava-Gregory's series. That method takes a long time to evaluate the value of π correctly to, say, 4 places of decimal. There is also the formula

$$\frac{\pi^4}{90} = \frac{1}{1^4} + \frac{1}{2^4} + \frac{1}{3^4} + \cdots + \frac{1}{n^4} \cdots .$$

To find the sum of the above series correct to, say, 4 places of decimal, it is enough to sum the first 25 terms. But the sum gives an approximate value of $\frac{\pi^4}{90}$. To get the value of π we need to multiply the sum by 90 and then extract its fourth root. Now the question is: how would the error committed in evaluating $\frac{\pi^4}{90}$ be propagated in the subsequent calculations? Put in a different way, let us write $x = \frac{\pi^4}{90}$ and $f(x) = \sqrt[4]{90x}$, and let x_n be the approximate value of x calculated by summing the first n terms of the series. So the approximate value of π would be $f(x_n)$. This leads naturally to the question: how is the error $|f(x) - f(x_n)|$ in the value of $f(x)$ related to the error $|x - x_n|$ in the value of x? Can we calculate $f(x)$ correct to the desired accuracy by calculating x sufficient accurately? If not, then this method of calculation is not very useful. If a small perturbation in x causes an abrupt change in $f(x)$, then we should perhaps think of some other method of calculation. It may be noted that, philosophically, continuity forms the basis of large parts of the experimental sciences where it is tacitly assumed that small errors in measurement will not lead to drastic changes in conclusions.

These ideas lead to the definition of continuity of functions.

DEFINITION: Let $f : \mathbb{R} \to \mathbb{R}$ be a function. We say that the function is *continuous* at $x_0 \in \mathbb{R}$ if for every $\varepsilon > 0$ we can find a $\delta > 0$ such that

$$|f(x) - f(x_0)| < \varepsilon \text{ whenever } |x - x_0| < \delta.$$

This is to say that for continuous functions f, the value of $f(x)$ can be restricted within the interval $(f(x_0) - \varepsilon,\ f(x_0) + \varepsilon)$ by restricting the value of x within $(x_0 - \delta, x_0 + \delta)$. (see Figure 3.1).

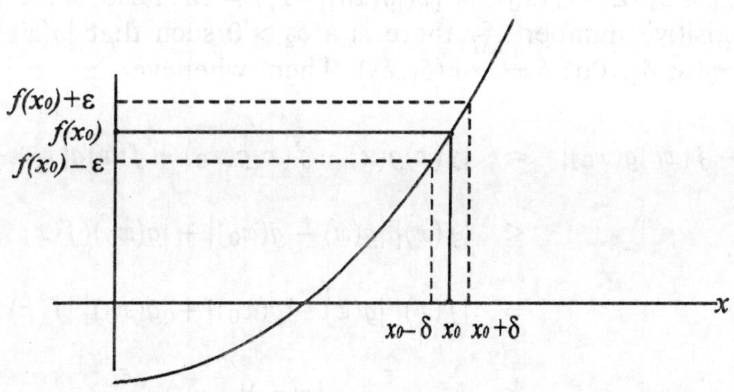

Figure 3.1

Note that for continuity at x_0, we need an interval containing x_0 to be contained in its domain, and hence it is enough that the function f has its domain an interval

$[a, b]$. So we can define continuity of a function $f : [a, b] \to \mathbb{R}$ in the same way as above. In the case of the end points a and b, we can only talk of the right and left continuity respectively.

Now that we have defined continuity at a point, we say that a function $f : [a, b] \to \mathbb{R}$ is a *continuous function* if it is continuous at every $x \in [a, b]$. We have then the following result.

THEOREM: *If $g, f : [a, b] \to \mathbb{R}$ are continuous functions and c is a constant, then*
(a) $f + g$ is a continuous function,
(b) $f - g$ is a continuous function,
(c) cf is a continuous function,
(d) fg is a continuous function,
where $(f + g)(x) = f(x) + g(x)$, $(f - g)(x) = f(x) - g(x)$, $(cf)(x) = cf(x)$ and $(fg)(x) = f(x)g(x)$.

PROOF: We shall only prove statement (d). Choose and fix any $\varepsilon > 0$. Since f is continuous at x_0, we have that for the positive number $\frac{\varepsilon}{2(|g(x_0)|+1)}$ there exists a $\delta_1 > 0$ such that

$$|f(x) - f(x_0)| < \frac{\varepsilon}{2(|g(x_0)| + 1)}$$

whenever $|x - x_0| < \delta_1$. Since $| \, |f(x)| - |f(x_0)|| \leq |f(x) - f(x_0)|$, we conclude that

$$|f(x)| < |f(x_0)| + \varepsilon/\{2(|g(x_0)| + 1)\}$$

whenever $|x - x_0| < \delta_1$. Let $|f(x_0)| + \varepsilon\{2(|g(x_0)| + 1)\} = M$. Also, since g is continuous at x_0, for the positive number $\frac{\varepsilon}{2M}$ there is a $\delta_2 > 0$ such that $|g(x) - g(x_0)| < \frac{\varepsilon}{2M}$ whenever $|x - x_0| < \delta_2$. Put $\delta = \min(\delta_1, \delta_2)$. Then, whenever $|x - x_0| < \delta$ we have

$$\begin{aligned}
|f(x)g(x) - f(x_0)g(x_0)| &= |f(x)g(x) - f(x)g(x_0) + f(x)g(x_0) - f(x_0)g(x_0)| \\
&\leq |f(x)| \, |g(x) - g(x_0)| + |g(x_0)(f(x) - f(x_0)| \\
&= |f(x)| \, |g(x) - g(x_0)| + |g(x_0)| \, |f(x) - f(x_0)| \\
&< M \cdot \frac{\varepsilon}{2M} + |g(x_0)| \cdot \frac{\varepsilon}{2(|g(x_0)| + 1)} \\
&= \frac{\varepsilon}{2} + \frac{\varepsilon}{2} \\
&= \varepsilon.
\end{aligned}$$

■

Observe that we have not claimed that the quotient of two continuous functions is continuous. The problem is obvious: $\frac{f(x)}{g(x)}$ cannot have any meaning at x for which $g(x) = 0$. So the question would be, if $g(x) \neq 0$ for every $x \in [a, b]$, is the function $h : [a, b] \to \mathbb{R}$, defined by $h(x) = f(x)/g(x)$, continuous? The answer is yes. For a proof, we need a preliminary result.

LEMMA: *If $g : [a, b] \to \mathbb{R}$ is continuous and $g(x_0) \neq 0$, then there is an $m > 0$ and $\delta > 0$ such that if $x_0 - \delta < x < x_0 + \delta$ then $|g(x)| > m$.*

PROOF: Let $|g(x_0)| = 2m$. Now $m > 0$. By continuity of g there is a $\delta > 0$ such that

$$|g(x) - g(x_0)| < m \quad \text{for} \quad x_0 - \delta < x < x_0 + \delta.$$

But $|g(x) - g(x_0)| \geq ||g(x)| - |g(x_0)||$ and hence $-m < |g(x)| - |g(x_0)| < m$, giving us

$$m = |g(x_0)| - m < |g(x)| \quad \text{for} \quad x_0 - \delta < x < x_0 + \delta. \qquad \blacksquare$$

The lemma says that if a continuous function does not vanish at a point, then there is an interval containing it in which it does not vanish at any point.

THEOREM: *If $f, g : [a, b] \to \mathbb{R}$ are continuous and $g(x) \neq 0$ for all x, then $h : [a, b] \to \mathbb{R}$ defined by $h(x) = \frac{f(x)}{g(x)}$ is continuous.*

The proof of this theorem (using the above lemma) is left as an exercise.

Examples

(a) $f : \mathbb{R} \to \mathbb{R}$ defined by $f(x) = a_0$ for all $x \in \mathbb{R}$, where $a_0 \in \mathbb{R}$ is continuous.

(b) $f : \mathbb{R} \to \mathbb{R}$ defined by $f(x) = x$ is continuous.

(c) $g : \mathbb{R} \to \mathbb{R}$ defined by $g(x) = x^2$ is a continuous function because $g(x) = f(x)f(x)$, where $f(x) = x$. Since f is continuous, by (b), g must be continuous.

(d) $h : \mathbb{R} \to \mathbb{R}$ defined by $h(x) = x^n$, n being a positive integer, is continuous by repeated application of the above reasoning.

(e) $p : \mathbb{R} \to \mathbb{R}$ defined by $p(x) = a_0 + a_1 x + \cdots + a_n x^n$, where $a_0, a_1, \cdots, a_n \in \mathbb{R}$, is also continuous. This is because of the fact that if $f_1, f_2, \cdots, f_n : \mathbb{R} \to \mathbb{R}$ are defined by $f_1(x) = x$, $f_2(x) = x^2, \cdots, f_n(x) = x^n$, then $a_1 f_1, a_2 f_2, \cdots, a_n f_n$ are also continuous functions. Hence,

$$a_0 + a_1 f_1 + \cdots + a_n f_n = p$$

is also a continuous function as the sum of continuous functions is a continuous function. Thus we have shown that a polynomial is a continuous function.

(f) Let p and q be polynomials. Let $\alpha_1, \cdots, \alpha_n \in \mathbb{R}$ be such that $q(\alpha_1) = q(\alpha_2) = \cdots = q(\alpha_n) = 0$ and $q(\alpha) \neq 0$ for $\alpha \neq \alpha_1$, $\alpha \neq \alpha_2, \cdots, \alpha \neq \alpha_n$. Let $D = \mathbb{R} - \{\alpha_1, \cdots, \alpha_n\}$. Then $h : D \to \mathbb{R}$ defined by $h(x) = \frac{p(x)}{q(x)}$ is a continuous function. What we have said is that a rational function which is defined everywhere except on the finite set of zeroes of the denominator is continuous.

(g) $f : \mathbb{R} \to \mathbb{R}$ defined by $f(x) = \sin x$ is continuous everywhere. Indeed $f(x) - f(x_0) = \sin x - \sin x_0 = 2 \sin \frac{x - x_0}{2} \cos \frac{x + x_0}{2}$. Therefore

$$|f(x) - f(x_0)| = 2\left|\sin \frac{(x - x_0)}{2}\right|\left|\cos \frac{x + x_0}{2}\right| \leq |x - x_0|$$

(since $|\sin \theta| \leq |\theta|$ when θ is measured in radians).

(h) $f : \mathbb{R} \to \mathbb{R}$ defined by $f(x) = \cos x$ is continuous since

$$|f(x) - f(x_0)| = |\cos x - \cos x_0| = 2\left|\sin \frac{x_0 - x}{2} \sin \frac{x + x_0}{2}\right| \leq |x - x_0|.$$

(i) $f : \mathbb{R} - \{(2n+1)\frac{\pi}{2} : n \in \mathbb{Z}\} \to \mathbb{R}$ defined by $f(x) = \tan x$ is continuous. We had to omit numbers like $\cdots, -\frac{3\pi}{2}, \frac{-\pi}{2}, \frac{\pi}{2}, \frac{3\pi}{2}, \cdots$ from the domain of f as $\tan x$ cannot be defined for these values of x.

(j) $f : \mathbb{R}_+ \to \mathbb{R}$ defined by $f(x) = x^{1/n}$ is a continuous function. Indeed

$$f(x) - f(a) = x^{1/n} - a^{1/n} = \frac{(x - a)}{x^{\frac{n-1}{n}} + x^{\frac{n-2}{n}} \frac{1}{a^n} + \cdots + a^{\frac{n-1}{n}}}.$$

Choose $|x - a| < |\frac{a}{2}|$ to start with, so that $|\frac{a}{2}| < |x| < \frac{3}{2}|a|$. Thus,

$$\left|x^{\frac{n-1}{n}} + x^{\frac{n-2}{n}} a^{\frac{1}{n}} + \cdots + a^{\frac{n-1}{n}}\right| > |a|^{\frac{n-1}{n}} \left(\left(\frac{1}{2}\right)^{\frac{n-1}{n}} + \left(\frac{1}{2}\right)^{\frac{n-2}{n}} + \cdots + 1\right).$$

Given an $\varepsilon > 0$, let $\delta = \min\left\{\frac{|a|}{2}, \varepsilon |a|^{\frac{n-1}{n}}\left(\left(\frac{1}{2}\right)^{\frac{n-1}{n}} + \cdots + 1\right)\right\}$. Then for $|x - a| < \delta$ we have

$$|f(x) - f(a)| = \frac{|x-a|}{\left|x^{\frac{n-1}{n}} + x^{\frac{n-2}{n}}a^{\frac{1}{n}} + \cdots + a^{\frac{n-1}{n}}\right|} < \varepsilon.$$

It can be shown that f defined by $f(x) = x^r$ is also continuous for every real $r \in \mathbb{R}$.

(k) Consider the function $f : \mathbb{R} \to \mathbb{R}$ defined by $f(x) = a^x$, where $a > 0$. Is f a continuous function? This is left as an exercise. (**Hint**: It will suffice to prove continuity at $x = 0$. This would follow from $\lim_{m\to\infty} a^{\frac{1}{m}} = 1$.)

We now discuss a few discontinuous functions.

(l) Recall the post office function discussed earlier. We have to pay Rs. 3 for mailing a letter weighing up to 15 grams. However, one pays Re 1 extra for every extra 10 grams or parts thereof. The post master charges you Re 1 extra the moment the weight of the mail exceeds 15 grams by a little amount. In fact $|f(15) - f(w)| \geq 1$ for $w > 15$, where $f(w)$ is the postage on the weight w. The inequality above tells us that the value of $f(w)$ changes abruptly as a consequence of any change in the value of w to the right, howsoever small it may be. So the post office function is discontinuous at 15, 25, 35,..., etc.

One can also easily see that $f : \mathbb{R} \to \mathbb{R}$ defined by $f(x) = [x]$ is discontinuous at every integral point. (Here $[x]$ is the integral part of the real number x. In the decimal representation of a real number as in Chapter 2, $[x]$ is the integer a_0, the first term).

(m) Suppose $f : \mathbb{R} \to \mathbb{R}$ is defined by $f(x) = \frac{1}{x}$ if $x \neq 0$ and $f(0) = 0$. We can see that f is not continuous at 0 as $f(x)$ changes abruptly when x goes over from negative to positive values.

We now turn to the question whether $f(x_n)$ approximates $f(x)$ when x_n approximates x. This is the same as asking: suppose $(x_n)_{n=1}^{\infty}$ is a sequence of real numbers converging to x, does $(f(x_n))_{n=1}^{\infty}$ converge to $f(x)$?

THEOREM: *$f : \mathbb{R} \to \mathbb{R}$ is continuous at $x \in \mathbb{R}$ if and only if $(f(x_n))_{n=1}^{\infty}$ converges to $f(x)$ whenever $(x_n)_{n=1}^{\infty}$ converges to x, i.e.,*

$$\lim_{n\to\infty} f(x_n) = f(\lim_{n\to\infty} x_n).$$

PROOF: Suppose f is continuous at x and $(x_n)_{n=1}^{\infty}$ converges to x. By continuity, for every $\varepsilon > 0$, there exists $\delta > 0$ such that $|f(x) - f(y)| < \varepsilon$ whenever $|x - y| < \delta$. Since $(x_n)_{n=1}^{\infty}$ converges to x, for this $\delta > 0$ we can find an n_0 such that

$$|x_n - x| < \delta \quad \text{for} \quad n > n_0.$$

So $|f(x) - f(x_n)| < \varepsilon$ for $n > n_0$ as $|x_n - x| < \delta$.

Conversely, suppose $(f(x_n))_{n=1}^{\infty}$ converges to $f(x)$ when $(x_n)_{n=1}^{\infty}$ converges to x. We have to show that f is continuous at x. Suppose f is not continuous at x. That is to say, there is an $\varepsilon > 0$ such that however small $\delta > 0$ we may choose, there will be a y satisfying $|x - y| < \delta$ yet $|f(x) - f(y)| \geq \varepsilon$. So for every n, let x_n be such a number for which $|x - x_n| < \frac{1}{n}$ and $|f(x) - f(x_n)| \geq \varepsilon$. Now we see that $(x_n)_{n=1}^{\infty}$ converges to x. But $(f(x_n))_{n=1}^{\infty}$ does not converge to $f(x)$ violating our hypothesis. So f must be continuous at x. ∎

Observe that the above theorem states that not only does $(f(x_n))_{n=1}^{\infty}$ converge whenever $(x_n)_{n=1}^{\infty}$ does, but also that $\lim f(x_n) = f(\lim x_n)$ for a continuous function. On the other hand, if f is not continuous it may happen that $\lim f(x_n)$ exists but does not equal $f(\lim x_n)$. This leads to another notion called *limit* of a function.

DEFINITION: Suppose $f : \mathbb{R} \to \mathbb{R}$. We say that ℓ is the limit of $f(x)$ as x tends to x_0 if for every $\varepsilon > 0$ there is a $\delta > 0$ such that

$$|f(x) - \ell| < \varepsilon \quad \text{for} \quad 0 < |x - x_0| < \delta.$$

In this case we write $\lim_{x \to x_0} f(x) = \ell$. This is the same thing as saying that $f(x)$ can be brought as close to ℓ as we please by bringing x sufficiently close to x_0. But we do not require the function to have the value ℓ at x_0. If it has the value ℓ at x_0, then it is continuous.

From the above discussions we see that for $\lim_{x \to x_0} f(x)$ to exist it is not necessary for us to assume that f is defined at x_0. We only need to know whether or not $f(x)$ is coming close to a definite real number when x is coming close to x_0. Thus for the limit of $f(x)$ (as x tends to x_0) to exist or not exist, we need the function f to be defined for $\{x : 0 < |x - x_0| < \delta\}$ for some $\delta > 0$. This is the set of points in the interval $(x_0 - \delta, x_0 + \delta)$ from which x_0 has been *deleted*, i.e., $\{x : 0 < |x - x_0| < \delta\} = (x_0 - \delta, x_0) \cup (x_0, x_0 + \delta)$. When we want to look at a function over such a punctured neighbourhood of a point, we try to see what happens to the function when we come close to it, without actually reaching the point. This is not really artificial at all as many physical and mathematical exigencies force us to look at such situations. Take the case of a particle moving in a straight line. With reference to a fixed point on the straight line, let $x(t)$ and $x(t_0)$ be the positions of the particle at time t and t_0 respectively ($t > t_0$). So its *average velocity* in the interval of time $[t_0, t]$ is given by $\frac{x(t)-x(t_0)}{t-t_0} = f(t)$. Now the function f is defined for every t *save* t_0. But the instantaneous velocity should indeed be the $\lim\limits_{t \to t_0} f(t)$. In other words, as the interval of time $[t_0, t]$ decreases, the average velocity should eventually stabilize to a certain number $v(t_0)$ called the instantaneous velocity of the particle at the instant of time t_0. If this happens, only then is it meaningful to talk of instantaneous velocity. See what we are doing. We have a function f defined for every real number except a particular real number t_0. Then we are trying to find out what happens to $f(t)$ as t comes closer and closer to t_0. We shall see later many such situations of finding the limit of $f(t)$ as $t \to t_0$, where $f(t)$ is defined for every t except $t = t_0$.

Let us look again at the post office function described earlier. This function is clearly defined for every x. But what happens to $f(x)$, for instance, when x comes close to 15? When we take $x = 14.9$, $x = 14.99$, $x = 14.999$ we are, at each successive stage, coming closer to 15. Similarly as we go through $x = 15.1, x = 15.01, x = 15.001$, we are coming closer to 15 as well. But in the former case we were approaching x through the values of x less than 15 or, what is the same thing as approaching 15 from the *left*. In the latter case we are approaching x through the values of x larger than 15 or we are approaching from the *right*. It is clear that $\lim f(x) = 2$ as x approaches 15 from the left, while $\lim f(x) = 3$ when x approaches 15 from the right. So we have the following definition.

DEFINITION: Let $f : \mathbb{R} - \{a\} \to \mathbb{R}$. We say that the left hand limit of $f(x)$ exists as x tends to a if there is a number ℓ such that for every $\varepsilon > 0$ we can find a $\delta > 0$ such that

$$|f(x) - \ell| < \varepsilon \quad \text{for} \quad a - \delta < x < a.$$

In such a case we call ℓ the left hand limit of $f(x)$ at a, and write $\lim\limits_{x \to a-} f(x) = \ell$. Similarly we say that $f(x)$ has a right hand limit r as x tends to a if for all $\varepsilon > 0$ there exists $\delta > 0$ such that

$$|f(x) - r| < \varepsilon \quad \text{for} \quad a < x < a + \delta.$$

In such a case we call r the right hand limit of $f(x)$ as x tends to a and write $\lim\limits_{x \to a+} f(x) = r$. It is clear that when the left hand limit exists it is unique. So is the case with the right hand limit. Thus we are led to the conclusion:

The limit of $f(x)$ exists as x tends to a if and only if both the left hand and right hand limits of $f(x)$ exist as x approaches a, and they are equal. Moreover, if the common *limit is equal to the value of the function at the point, then the function is continuous at the point.*

In the case of the post office function, the left hand and right hand limits exist at 15 but they are *not* equal. So there is no question of the limit existing, much less the continuity of the function at 15.

Some Elementary Properties of Limits

(1) Suppose for functions f and g, $\lim_{x \to a} f(x)$ and $\lim_{x \to a} g(x)$ exist. Then we have

(a) $\lim_{x \to a} (f(x) \pm g(x)) = \lim_{x \to a} f(x) \pm \lim_{x \to a} g(x)$.

(b) $\lim_{x \to a} (f(x)\, g(x)) = \lim_{x \to a} f(x) \lim_{x \to a} g(x)$.

(c) $\lim_{x \to a} \dfrac{f(x)}{g(x)} = \dfrac{\lim f(x)}{\lim g(x)}$ if $\lim_{x \to a} g(x) \neq 0$.

(2) Suppose f is continuous at a, then $\lim_{x \to a} f(x)$ is simply $f(a)$.

(3) Suppose f is defined in a punctured neighbourhood of x_0, g is defined in a neighbourhood of x_0 and is continuous. If $f(x) = g(x)$ for x in the punctured neighbourhood of x_0, then $\lim_{x \to x_0} f(x) = g(x_0)$.

From the above, it is easy to see that

(a) $\lim_{x \to x_0} p(x) = p(x_0)$ for a polynomial p,

(b) $\lim_{x \to x_0} \sin x = \sin x_0$,

(c) $\lim_{x \to x_0} \dfrac{p(x)}{q(x)} = \dfrac{p(x_0)}{q(x_0)}$ for polynomials p, q, if $q(x_0) \neq 0$.

Before we move on to the evaluation of other limits, we prove the following useful result:

THEOREM: *Suppose $f : (a, b) \to \mathbb{R}$ is continuous at $c \in (a, b)$. Suppose $g : I \to (a, b)$, where I is an open interval and $x_0 \in I$. If $\lim_{x \to x_0} g(x)$ exists and is equal to c, then*

$$\lim_{x \to x_0} f(g(x)) = f(c).$$

PROOF: Since f is continuous at x_0, for each $\varepsilon > 0$ we can find $\delta' > 0$ such that

$$|f(y) - f(c)| < \varepsilon \quad \text{for} \quad |y - c| < \delta'.$$

Since $\lim_{x \to x_0} g(x) = c$, for the given δ' we can find a $\delta > 0$ such that

$$|g(x) - c| < \delta' \quad \text{for} \quad 0 < |x - x_0| < \delta.$$

So for $0 < |x - x_0| < \delta$ we have

$$|f(g(x)) - f(c)| < \varepsilon \quad \text{since} \quad |g(x) - c| < \delta'.$$

■

COROLLARY: *Let I and J be open intervals, $g : I \to J$ be continuous at $x_0 \in I$. If $f : J \to \mathbb{R}$ is continuous at $g(x_0) \in J$, then $f \circ g : I \to \mathbb{R}$ is continuous at x_0. This means that the composition of continuous functions is continuous.*

Exercise: If f is continuous, show that $|f|$ is continuous (where $|f|(x) = |f(x)|$).

Some Useful Limits

(1) (a) $\lim_{x \to a} \frac{x^n - a^n}{x - a} = na^{n-1}$ for a positive integer n. For this, note that $\frac{x^n - a^n}{x - a} = x^{n-1} + x^{n-2}a + \cdots + xa^{n-2} + a^{n-1}$ for $x \neq a$, and hence

$$\lim_{x \to a} \frac{x^n - a^n}{x - a} = \lim_{x \to a} (x^{n-1} + x^{n-2}a + \cdots + xa^{n-2} + a^{n-1})$$
$$= na^{n-1}.$$

(b) $\lim_{x \to a} \frac{x^n - a^n}{x - a} = na^{n-1}$ for a negative integer n if $a \neq 0$.

If n is a negative integer write $n = -m$, where $m > 0$.

$$\lim_{x \to a} \frac{x^n - a^n}{x - a} = \lim_{x \to a} \frac{x^{-m} - a^{-m}}{x - a} = \lim_{x \to a} \frac{a^m - x^m}{x - a} \cdot \frac{1}{x^m a^m}$$
$$= -ma^{m-1} \cdot \frac{1}{a^{(2m)}} = -ma^{-m-1} = na^{n-1}.$$

(c) $\lim_{x \to a} \frac{x^{\frac{p}{q}} - a^{\frac{p}{q}}}{x - a} = \frac{p}{q} a^{\frac{p}{q} - 1}, p, q \in \mathbb{Z}, q \neq 0$ and when $a > 0$.

Write $y = x^{1/q}$, $b = a^{1/q}$. So we have $x = y^q, a = b^q$

$$\lim_{x\to a}\frac{x^{\frac{p}{q}}-a^{\frac{p}{q}}}{x-a} = \lim_{y\to b}\frac{y^p-b^p}{y^q-b^q} = \lim_{y\to b}\frac{\frac{y^p-b^p}{y-b}}{\frac{y^q-b^q}{y-b}} = \frac{pb^{p-1}}{qb^{q-1}}$$

$$= \frac{p}{q}b^{p-q} = \frac{p}{q}a^{\frac{p-q}{q}} = \frac{p}{q}a^{\frac{p}{q}-1}.$$

(2) $\lim_{\theta\to 0}\frac{\sin\theta}{\theta} = 1$, θ measured in radians.

We first observe that $\frac{\sin\theta}{\theta} = \frac{\sin(-\theta)}{-\theta}$. So it suffices to show that $\lim_{\theta\to 0+}\frac{\sin\theta}{\theta} = 1$.

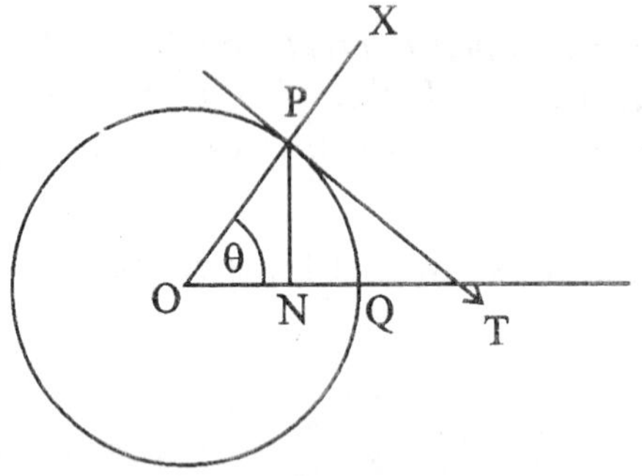

Figure 3.2

Consider Figure 3.2. Draw a line and mark points O and Q on it. Draw a line OX through O making an angle θ to OQ. Draw a circle of radius equal to the lengh OQ. Let it meet OX at P. Draw a tangent to the circle at P. Let it meet extended OQ at T. Drop a perpendicular from P on OQ. Let it meet OQ at N.

Now it is clear that the area of the triangle OPN is less than the area of the sector OPQ which is less than the area of the triangle OPT. The areas of the triangle OPN, the sector OPQ and the triangle OPT are $\frac{1}{2}r^2\cos\theta\sin\theta$, $\frac{1}{2}r^2\theta$ and $\frac{1}{2}r^2\tan\theta$ respectively (where r is the length OP and the angle θ is measured in radians). Thus, $\frac{1}{2}r^2\sin\theta\cos\theta < \frac{1}{2}r^2\theta < \frac{1}{2}r^2\tan\theta$, which gives $\sin 2\theta < 2\theta$ and $\tan\theta > \theta$, leading to $\cos\theta < \frac{\sin\theta}{\theta} < 1$. Thus $0 < 1-\frac{\sin\theta}{\theta} < 1-cos\,\theta = 2\sin^2\frac{\theta}{2} = 2(\sin\frac{\theta}{2})^2 < \frac{\theta^2}{2}$. So for $\varepsilon > 0$ choose $\delta = \sqrt{2\varepsilon}$. Now we have for $|\theta| < \delta$,

$$\left|1-\frac{\sin\theta}{\theta}\right| < \frac{\theta^2}{2} < \varepsilon.$$

In some cases, when the function $f:\mathbb{R}\to\mathbb{R}$ is defined and there is an ℓ such that for every $\varepsilon > 0$ we have an x_0 such that

$$|f(x)-\ell| < \varepsilon \quad \text{for} \quad x > x_0,$$

we say that $f(x)$ *tends to* ℓ as x *tends to* ∞. Here the notion of limit looks like the convergence of sequences. We write

$$\lim_{x\to\infty} f(x) = \ell.$$

Be warned that ∞ is not a real number to which x is coming close. Similarly, we may define $\lim\limits_{x\to-\infty} f(x) = \ell$. If for every $M > 0$ we can find a $\delta > 0$ such that $f(x) > M$ for $0 < |x - a| < \delta$, then we can write $\lim\limits_{x\to a} f(x) = \infty$. A similar definition for $\lim\limits_{x\to a} f(x) = -\infty$ can be given.

$$\text{(3) (a)} \lim_{x\to\infty}\left(1+\frac{1}{x}\right)^x = e. \quad \text{(b)} \lim_{x\to-\infty}\left(1+\frac{1}{x}\right)^x = e.$$

PROOF: (a) Let $n \le x < n+1$ so that $1 + \frac{1}{n+1} < 1 + \frac{1}{x} \le 1 + \frac{1}{n}$, giving us $(1+\frac{1}{n+1})^n < (1+\frac{1}{x})^x \le (1+\frac{1}{n})^{n+1}$. This implies that $(1+\frac{1}{n+1})^{n+1}(1+\frac{1}{n+1})^{-1} < (1+\frac{1}{x})^x \le (1+\frac{1}{n})^n(1+\frac{1}{n})$.

Now as $x \to \infty, n$ also tends to ∞ and in such a case both the right hand and left hand side of the above inequality tend to e. So $\lim\limits_{x\to\infty} (1+\frac{1}{x})^x = e$.

(b) In the case of $x \to -\infty$, let us write $y = -x$ so that

$$\begin{aligned}\left(1+\frac{1}{x}\right)^x &= \left(1-\frac{1}{y}\right)^{-y} = \frac{1}{\left(\dfrac{y-1}{y}\right)^y} \\ &= \left(\frac{y}{y-1}\right)^y = \left(1+\frac{1}{y-1}\right)^{y-1}\left(1+\frac{1}{y-1}\right).\end{aligned}$$

Now $\lim\limits_{y\to\infty}(1+\frac{1}{y-1})^{y-1} = e$. So we have

$$\lim_{x\to-\infty}\left(1+\frac{1}{x}\right)^x = \lim_{y\to\infty}\left(1-\frac{1}{y}\right)^{-y} = \lim\left(1+\frac{1}{y-1}\right)^{y-1}\left(1+\frac{1}{y-1}\right) = e. \quad \blacksquare$$

COROLLARY: $\lim\limits_{x\to\infty}\left(1-\dfrac{1}{x}\right)^x = \dfrac{1}{e}$.

PROOF

$$\begin{aligned}\lim_{x\to\infty}\left(1-\frac{1}{x}\right)^x &= \lim_{x\to\infty}\left(\frac{x-1}{x}\right)^x = \lim_{x\to\infty}\frac{1}{\left(\dfrac{x}{x-1}\right)^x} \\ &= \lim_{x\to-\infty}\frac{1}{\left(1+\dfrac{1}{x-1}\right)^{x-1}} \cdot \frac{1}{1+\dfrac{1}{x-1}} = \frac{1}{e}.\end{aligned}$$

$\blacksquare$

COROLLARY: $\lim_{n\to\infty}\left(1+\frac{x}{n}\right)^n = e^x$.

PROOF: Writing $u = \frac{n}{x}$, note that

$$\lim_{n\to\infty}\left(1+\frac{x}{n}\right)^n = \lim_{u\to\infty}\left(\left(1+\frac{1}{u}\right)^u\right)^x = e^x. \qquad \blacksquare$$

COROLLARY: $\lim_{t\to 0}(1+t)^{1/t} = e$.

It will be appropriate to mention something about left continuity and right continuity of functions.

DEFINITION: If $\lim_{x\to x_0-} f(x) = f(x_0)$, then we say that the function f is *left continuous* at x_0. If $\lim_{x\to x_0+} f(x) = f(x_0)$, then we say that the function is *right continuous*.

Thus a function may be left continuous but not right continuous, in which case the function cannot be continuous. It may be right continuous but not left continuous, in which case also the function cannot be continuous. If a function is both left continuous and right continuous, only then is it continuous.

Some Further Properties of Continuous Functions

INTERMEDIATE VALUE THEOREM: *Let $f : [a,b] \to \mathbb{R}$ be continuous. Suppose for $x_1, x_2 \in [a,b]$, $f(x_1) \neq f(x_2)$. If c is a real number between $f(x_1)$ and $f(x_2)$, then there is an x_0 between x_1 and x_2 such that $f(x_0) = c$.*

PROOF: Define $g : [a,b] \to \mathbb{R}$ by $g(x) = f(x) - c$. Then g is a continuous function such that $g(x_1)$ and $g(x_2)$ are of opposite sign. The assertion of the theorem amounts to saying that there is a point x_0 between x_1 and x_2 such that $g(x_0) = 0$. Without loss of generality we may take $g(x_1) > 0$ and $g(x_2) < 0$ (otherwise replace g by $-g$). If $g(\frac{x_1+x_2}{2}) = 0$, then $x_0 = \frac{x_1+x_2}{2}$. If $g(\frac{x_1+x_2}{2}) > 0$, write $\frac{x_1+x_2}{2} = a_1$ and $x_2 = b_1$; otherwise, write $a_1 = x_1$ and $b_1 = \frac{x_1+x_2}{2}$ so that we have $g(a_1) > 0$, $g(b_1) < 0$. Now if $g(\frac{a_1+b_1}{2}) = 0$, then $x_0 = \frac{a_1+b_1}{2}$. If $g(\frac{a_1+b_1}{2}) > 0$, write $\frac{a_1+b_1}{2} = a_2$ and $b_1 = b_2$, otherwise write $a_2 = a_1$ and $b_2 = \frac{a_1+b_1}{2}$, so that we have $g(a_2) > 0$, $g(b_2) < 0$. We could continue this process and find sequences $(a_n)_{n=1}^{\infty}, (b_n)_{n=1}^{\infty}$ with $g(a_n) > 0$, $g(b_n) < 0$ and

$$a_1 \le a_2 \le \cdots \le a_n \le a_{n+1} \le \cdots \le x_2,$$

$$b_1 \geq b_2 \geq \cdots \geq b_n \geq b_{n+1} \geq \cdots \geq x_1.$$

Since $(a_n)_{n=1}^{\infty}$ is a monotonically non-decreasing sequence bounded above, it must converge. Suppose it converges to α. Similarly $(b_n)_{n=1}^{\infty}$ is monotonically non-increasing, bounded below and therefore converges to, say, β. We further note that $b_n - a_n = \frac{x_2 - x_1}{2^n} \to 0$ as $n \to \infty$ implying $\alpha = \beta$. Let us call this x_0. By the continuity of g we have $\lim_{n\to\infty} g(a_n) = g(x_0) = \lim_{n\to\infty} g(b_n)$, and since $g(a_n) > 0$ for all n, we must have $g(x_0) \geq 0$ and at the same time since $g(b_n) < 0$ for all n, we must also have $g(x_0) \leq 0$. This implies $g(x_0) = 0$. ■

COROLLARY: *If f is a continuous function in an interval I and $f(a)f(b) < 0$ for some $a, b \in I$, then there is a point c between a and b for which $f(c) = 0$.* **(Exercise).**

The above result is often used to locate the roots of equations of the form $f(x) = 0$.

For example, consider the equation

$$f(x) \equiv x^3 + x - 1 = 0.$$

Note that $f(0) = -1$ whereas $f(1) = 1$. This shows that the above equation has a root between 0 and 1. Now try with .5. $f(.5) = -0.375$. So there must be a root of the equation between 1 and 0.5. Try 0.75. $f(.75) > 0$, which means that the root is between 0.5 and 0.75. So we may try 0.625. $f(0.625) < 0$. So the root is between 0.75 and 0.625. Now if we take the approximate root to be .6875, then we are away from the exact root by at most a distance of .0625. If we continue this process further, we shall get better and better approximations to the root of the equation.

Exercise: Find the cube root of 10 using the above method correct to 4 places of decimal.

Exponentials and Logarithms

For $a > 1$ define $f : \mathbb{R} \to \mathbb{R}$ by $f(x) = a^x$. It is easily seen that $f(x) > f(y)$ if $x > y$. This shows that f is one-to-one. Further, $\lim_{x\to\infty} f(x) = \infty$, whereas $\lim_{x\to-\infty} f(x) = 0$. That $f : \mathbb{R} \to \mathbb{R}$ is onto $\mathbb{R}_+$ follows from the intermediate value theorem. Thus $f : \mathbb{R} \to \mathbb{R}_+$ defined by $f(x) = a^x$ is bijective. So there is a unique map

$$g : \mathbb{R}_+ \to \mathbb{R}$$

such that

$$f(g(y)) = y \text{ for every } y \text{ in } \mathbb{R}_+$$

and

$$g(f(x)) = x \quad \text{for every} \quad x \text{ in } \mathbb{R}.$$

This function g is what we call the logarithm function of y to the base a, written $\log_a y$. In fact, the logarithm is a continuous function.

For $y_0 \in \mathbb{R}_+$, $\varepsilon > 0$ let $\delta = \min\{a^{x_0+\varepsilon} - a^{x_0}, a^{x_0} - a^{x_0-\varepsilon}\}$, where $x_0 = \log_a y_0$. Then we have for $|y_0 - y| < \delta$, $a^{x_0-\varepsilon} \leq y_0 - \delta < y < y_0 + \delta \leq a^{x_0+\varepsilon}$, or

$$x_0 - \varepsilon < \log_a y < x_0 + \varepsilon \quad \text{or} \quad g(y_0) - \varepsilon < g(y) < g(y_0) + \varepsilon.$$

Exercise: If $f : \mathbb{R} \to \mathbb{R}$ is an increasing continuous function, show that it is bijective onto its range and its inverse is also continuous.

With the help of the logarithm function we can evaluate $\lim\limits_{x \to 0} \dfrac{a^x - 1}{x}$. Let $a^x = 1 + y$ so that $y \to 0$ as $x \to 0$. Also, $x = \log_a(1 + y)$. So we have

$$\begin{aligned} \lim_{x \to 0} \frac{a^x - 1}{x} &= \lim_{y \to 0} \frac{y}{\log_a(1 + y)} \\ &= \lim_{y \to 0} \frac{1}{\frac{1}{y} \log_a(1 + y)} = \lim_{y \to 0} \frac{1}{\log_a(1 + y)^{\frac{1}{y}}} \\ &= \frac{1}{\log_a e} = \log_e a. \end{aligned}$$

In the step before last, we have used the fact that the logarithm is a continuous function and that $\lim\limits_{y \to 0}(1 + y)^{1/y} = e$, while in the last step we have observed that $(\log_a e)^{-1} = \log_e a$. **(Exercise)**.

Boundedness of a Continuous Function

Suppose $f : I \to \mathbb{R}$ is a continuous function (where I is an interval). Now for every $x_0 \in I$ and $\varepsilon > 0$ we have a $\delta > 0$ such that $f(x_0) - \varepsilon < f(x) < f(x_0) + \varepsilon$ for $x_0 - \delta < x < x_0 + \delta$. This tells us that f is bounded in the interval $(x_0 - \delta,\ x_0 + \delta)$. Does it mean that the function is bounded in its entire domain? What we have shown is that given an $x \in I$, there is an interval I_x and two real numbers m_x and M_x such that

$$m_x < f(\xi) < M_x \quad \text{for all,} \quad \xi \in I_x.$$

Surely $\cup_{x\in I} I_x \supset I$. But if we could choose finitely many intervals out of the collection $\{I_x\}_{x\in I}$, say, $I_{x_1}, I_{x_2}, \cdots, I_{x_n}$ such that $I_{x_1} \cup I_{x_2} \cup \cdots \cup I_{x_n} \supset I$, then we would get $m < f(\xi) < M$, where $M = \max\{M_{x_1}, \cdots, M_{x_n}\}$ and $m = \min\{m_{x_1}, m_{x_2}, \cdots, m_{x_n}\}$. That we can indeed make such a choice is a property of a closed bounded interval I in $\mathbb{R}$ and is given by the following theorem, the proof of which, as given below, may be omitted in a first reading.

THEOREM (HEINE-BOREL): *Let $a, b \in \mathbb{R}$ and let $\mathcal{I}$ be a family of open intervals covering $[a, b]$, i.e., for all $x \in [a, b]$, there exists $I \in \mathcal{I}$ such that $x \in I$. Then we can find finitely many open intervals $I_1, \cdots, I_n \in \mathcal{I}$ such that $I_1 \cup I_2 \cup I_3 \cup \cdots \cup I_n \supset [a, b]$.*

PROOF: Suppose our contention is false. Let us take the intervals $[a, c]$ and $[c, b]$, where $c = \frac{a+b}{2}$. If the hypothesis is false, then it should be false for at least one of the intervals $[a, c]$ or $[c, b]$. Otherwise, we could find $I_1, \cdots, I_m \in \mathcal{I}$ and $J_1, \cdots, J_n \in \mathcal{I}$ such that $I_1 \cup \cdots \cup I_m \supset [a, c]$ and $J_1 \cup J_2 \cup \cdots \cup J_n \supset [c, b]$, and then $\{I_1, \cdots, I_m, J_1, \cdots, J_n\}$ would be the finite family of intervals for which $I_1 \cup I_2 \cup \cdots \cup I_m \cup J_1 \cup \cdots \cup J_n \supset [a, b]$.

So let us assume that at least for one of the intervals $[a, c]$ or $[c, b]$ the assumption of the theorem is false. Call it $[a_1, b_1]$. Again let $c_1 = \frac{a_1+b_1}{2}$. Now since the claim of the theorem is false for $[a_1, b_1]$ it should be false at least for $[a_1, c_1]$ or $[c_1, b_1]$ by the above argument. Call it $[a_2, b_2]$. We have $a \leq a_1 \leq a_2 < b_2 \leq b_1 \leq b$. We can continue this process to get a sequence of intervals $[a_1, b_1] \supset [a_2, b_2] \supset [a_3, b_3] \supset \cdots \supset [a_n, b_n] \supset \cdots$ for which the assertion is false. Observe further that $b_n - a_n = \frac{b-a}{2^n}$ and that we have $a \leq a_1 \leq a_2 \leq \cdots \leq a_n < b_n \leq b_{n-1} \leq \cdots \leq b$.

This gives us a monotonically increasing sequence $(a_n)_{n=1}^{\infty}$ which is bounded above and a monotonically decreasing sequence $(b_n)_{n=1}^{\infty}$ bounded below. So $(a_n)_{n=1}^{\infty}$ and $(b_n)_{n=1}^{\infty}$ must converge to say α and β respectively. Then $\alpha = \beta$ because $\beta - \alpha = \lim(b_n - a_n) = \lim\limits_{n\to\infty} \frac{b-a}{2^n} = 0$. Since $\mathcal{I}$ covers $[a, b]$, α must belong to J for some $J \in \mathcal{I}$. Also since $\lim\limits_{n\to\infty} a_n = \alpha$, there exists an n_1 such that $a_n \in J$, for all $n > n_1$, and similarly there exists an n_2 such that $b_n \in J$ for all $n > n_2$. Now let $n_0 = \max\{n_1, n_2\}$. Therefore we conclude that $[a_n, b_n] \subset J$ for all $n > n_0$. But this violates our hypothesis that we cannot choose finitely many members of $\mathcal{I}$ whose union will contain $[a_n, b_n]$ for any n. ∎

COROLLARY: *A continuous function on a closed interval $[a, b]$ is bounded.*

The proof of the corollary is already given just before the Heine-Borel theorem. So, if we have a continuous function $f : [a, b] \to \mathbb{R}$ and $M = \sup\{f(x) : a \leq x \leq b\}$ and $m = \inf\{f(x) : a \leq x \leq b\}$, the above corollary says $-\infty < m \leq M < \infty$. Next we ask the natural question: do there exist two points $x_0, y_0 \in [a, b]$ such that $f(x_0) = M$ and $f(y_0) = m$? In other words, does a continuous function on a closed interval attain its bounds? The answer is yes.

THEOREM: *Suppose $f : [a, b] \to \mathbb{R}$ is continuous, and $M = \sup\{f(x) : a \le x \le b\}$ and $m = \inf\{f(x) : a \le x \le b\}$. Then there are two points $x_0, y_0 \in [a, b]$ such that*

$$f(x_0) = M \text{ and } f(y_0) = m.$$

(Note: These points x_0 and y_0 need not be unique.)

PROOF: Suppose there is no point $x \in [a, b]$ such that $f(x) = M$, then we would have $f(x) < M$ or $M - f(x) > 0$ for all $x \in [a, b]$. Let us define $g : [a, b] \to \mathbb{R}$ by

$$g(x) = \frac{1}{M - f(x)}.$$

Since $M - f(x)$ vanishes nowhere, g is also a continuous function. So by the corollary above it ought to be bounded above and below. Let $0 < g(x) < M_1$, for all $x \in [a, b]$. On the other hand, by the property of a supremum, we note that there exists an $x \in [a, b]$ such that $f(x) + \frac{1}{2M_1} > M$, which implies that $M - f(x) < \frac{1}{2M_1}$ or $g(x) = \frac{1}{M - f(x)} > 2M_1$, which is a contradiction. Therefore, $f(x)$ must attain the value M at some point $x_0 \in [a, b]$. The proof of the other part is very similar. ■

The above theorem together with the corollary says that on a closed interval, a continuous function is bounded and attains its bounds. This, again by the intermediate value theorem, means that the function must attain all the values between its supremum and infimum. Thus the image of a closed interval under a continuous map is a closed interval. However, if f is a continuous map on an open interval, then the function need not be bounded.

Example: Let $f : (0, 1) \to \mathbb{R}$ be defined by $f(x) = \frac{1}{x}$. This is surely continuous but $\lim_{x \to 0} f(x) = \infty$, which means that given any $M > 0$, we can always find x such that $f(x) > M$, viz., choose $0 < x < \frac{1}{M}$.

If f is a continuous function, then given $\varepsilon > 0$, for each x_0 fixed, we can find $\delta > 0$ such that

$$|f(x) - f(x_0)| < \varepsilon \text{ whenever } |x - x_0| < \delta.$$

Here δ depends upon x_0.

Can we choose $\delta_0 > 0$ such that it works for all x_0? The answer in general is no.

Example: Let $f : \mathbb{R} \to \mathbb{R}$ be defined by $f(x) = x^2$. If we fix any $\theta > 0$, then for $x > 0$, $f(x + \theta) - f(x) = 2\theta x + \theta^2 \ge 2\theta x$, and hence as x becomes large, the difference between $f(x + \theta)$ and $f(x)$ also becomes large for every fixed $\theta > 0$. So for say $\varepsilon = 1$, we cannot choose $\delta > 0$ such that $|f(x+\theta) - f(x)| < \varepsilon$ for all $\theta < \delta$, and all x. We thus have the following definition:

DEFINITION: Let $f : D \to \mathbb{R}$ be a continuous function where $D = \mathbb{R}$ or $[a, b]$ or (a, b). Then f is said to be *uniformly continuous* if for all $\varepsilon > 0$, there exists a $\delta > 0$ such that

$$|f(x) - f(y)| < \varepsilon \text{ for all } x, y \in D \text{ with } |x - y| < \delta.$$

We have seen above that every continuous function need not be uniformly continuous. When $D = [a, b]$, however, every continuous function is uniformly continuous as the next result shows.

THEOREM: *Let $f : [a, b] \to \mathbb{R}$ be continuous. Then f is uniformly continuous.*

PROOF: Fix $\varepsilon > 0$. The continuity of f implies that for every $x \in [a, b]$, we can choose $\delta_x > 0$ such that

$$|f(x) - f(y)| < \varepsilon/2 \text{ whenever } |y - x| < \delta_x,\ y \in [a, b].$$

Now let $I_x = \left(x - \frac{1}{2}\delta_x, x + \frac{1}{2}\delta_x\right)$.

Then clearly $\{I_x : x \in [a, b]\}$ covers $[a, b]$ as $x \in I_x$. By the Heine-Borel theorem, we can get finitely many intervals out of this family, $I_{x_1}, I_{x_2}, \cdots, I_{x_m}$, such that

$$I_{x_1} \cup I_{x_2} \cup \cdots \cup I_{x_m} \supset [a, b].$$

Let $\delta = \min\{\frac{1}{2}\delta_{x_1}, \frac{1}{2}\delta_{x_2}, \cdots, \frac{1}{2}\delta_{x_m}\}$. Then $\delta > 0$ (note that the minimum of finitely many positive numbers is always positive). Next we claim that if $x, y \in [a, b]$, $|x - y| < \delta$, then $|f(x) - f(y)| < \varepsilon$.

Since $x \in [a, b] \subseteq I_{x_1} \cup \cdots \cup I_{x_m}$, we can find $k \leq m$ such that $x \in I_{x_k}$, i.e., $|x - x_k| < \frac{1}{2}\delta_{x_k}$. Now $|y - x_k| \leq |x - y| + |x - x_k| \leq \delta + \frac{1}{2}\delta_{x_k} \leq \delta_{x_k}$. Hence $|f(y) - f(x_k)| < \frac{\varepsilon}{2}$ and $|f(x) - f(x_k)| < \frac{\varepsilon}{2}$ and therefore $|f(y) - f(x)| < \varepsilon$. ■

3.2 Differentiation

We have seen how the concept of continuity is naturally associated with attempts to model gradual changes. For example, consider the linear function $f : \mathbb{R} \to \mathbb{R}$ given by $f(x) = ax + b$, where change in $f(x)$ is proportional to the change in x. This simple looking function is often used to model many practical problems. One such case is given below.

Suppose 30 men working for 7 hours a day can complete a piece of work in 16 days. In how many days can 28 men working for 6 hours a day complete the same work? It must be evident to most of the readers that the answer is $\frac{16\times7\times30}{28\times6} = 20$ days.

(While solving this we have tacitly assumed that the amount of work done is proportional to the number of men working, to the number of hours each man works per day and also to the number of days they work. Similarly, Charles' law for ideal gases states that the pressure remaining constant, the increase in volume of a mass of gas is proportional to the increase in temperature of the gas).

But there are exceptions to this as well. Galileo discovered that the distance covered by a body, falling from rest, is proportional to the square of the time for which it has fallen, and the velocity is proportional to the square root of the distance through which it has fallen. Similarly, Kepler's law tells us that the square of the period of a planet going round the sun is proportional to the cube of its mean distance from the sun.

These and many other problems involve functions that are not linear. If for example we plot the graph of the distance covered by a particle versus time, it is a straight line only when the motion is uniform. But we are seldom lucky to encounter only uniform motion. (Besides, uniform motion would be so monotonous. Perhaps there would be no life at all if all motions were uniform. Imagine a situation in which each body is in uniform motion. A body at rest would be eternally at rest and those once in motion, would never stop.) So the simple method of proportionality becomes quite inadequate to tackle such non-linear problems. The genius of Newton lay in looking at those problems which are next best to linear, the ones that are *nearly linear.*

We know that the graph of a linear function is a straight line. What Newton suggested was to look at functions, small portions of whose graphs look almost like a straight line (see Figure 3.3).

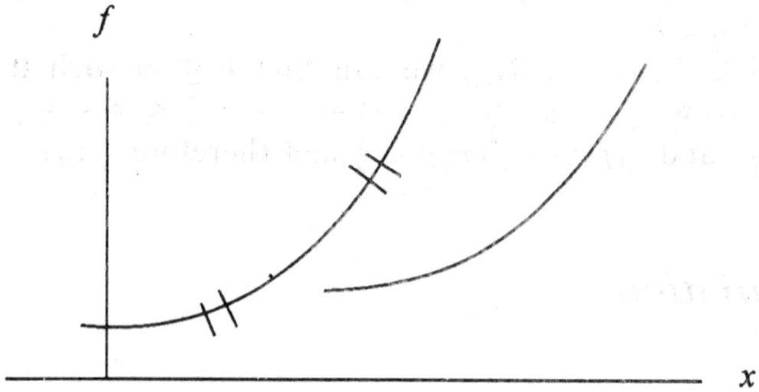

Figure 3.3

In Figure 3.3, the graph certainly is not a straight line. But a small portion of it looks like a part of a straight line. To formalize this idea, we need the concept of differentiability.

DEFINITION: Let I be an open interval and $f : I \to \mathbb{R}$ be a function. We say that f is locally linear or *differentiable* at $x_0 \in I$ if there is a constant m such that

$$f(x) - f(x_0) = m(x - x_0) + r(x_0, x)(x - x_0),$$

or equivalently, for x in a punctured interval around x_0,

$$\frac{f(x) - f(x_0)}{x - x_0} = m + r(x_0, x),$$

where $r(x_0, x) \to 0$ as $x \to x_0$. What this means is that for small enough $x - x_0$, $\frac{f(x)-f(x_0)}{x-x_0}$ is nearly a constant or, equivalently, $f(x) - f(x_0)$ is nearly proportional to the increment $x - x_0$. This is what is called the *principle of proportional parts* and used very often in calculations using tables, when the number for which we are looking up the table is not found there.

Thus if a function f is differentiable at x_0, then $\lim\limits_{x \to x_0} \frac{f(x)-f(x_0)}{x-x_0}$ exists and is called the *derivative* of f at x_0 and denoted by $f'(x_0)$. So we write

$$\lim_{x \to x_0} \frac{f(x) - f(x_0)}{x - x_0} = f'(x_0).$$

We need to look at functions which are not differentiable at some point, to fix our ideas. For example, consider the function $f : \mathbb{R} \to \mathbb{R}$ defined by $f(x) = |x|$.

This function though continuous at every point, is *not differentiable* at $x = 0$. In fact, $\lim\limits_{x \to 0_+} \frac{|x|}{x} = 1$, whereas $\lim\limits_{x \to 0-} \frac{|x|}{x} = -1$. What all this means is that if one looks at the graph of $f(x) = |x|$, it has a sharp corner at the origin (Figure 3.4).

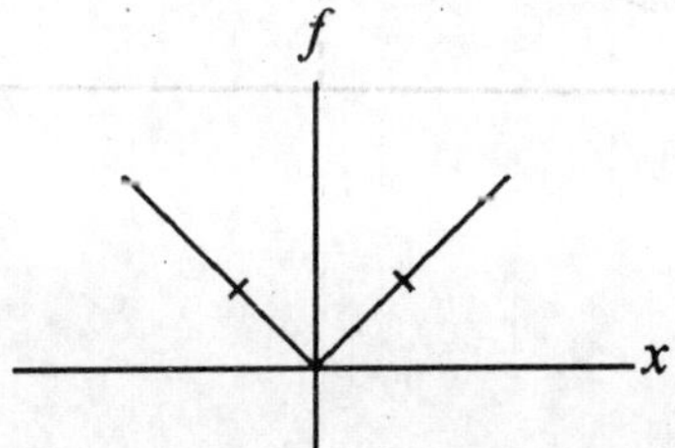

Figure 3.4

No matter how small a part of the graph containing the point (0,0) is taken, it *never* looks like a line segment. The reader can test for the non-differentiability of $f(x) = |\sin x|$ at $x = n\pi$.

This leads us to the notion of direction of the graph at a point: Suppose $f : I \to \mathbb{R}$ is a function differentiable at $x_0 \in I$, and let P and Q be the points $(x_0, f(x_0))$ and $(x, f(x))$ respectively in the graph of f (see Figure 3.5).

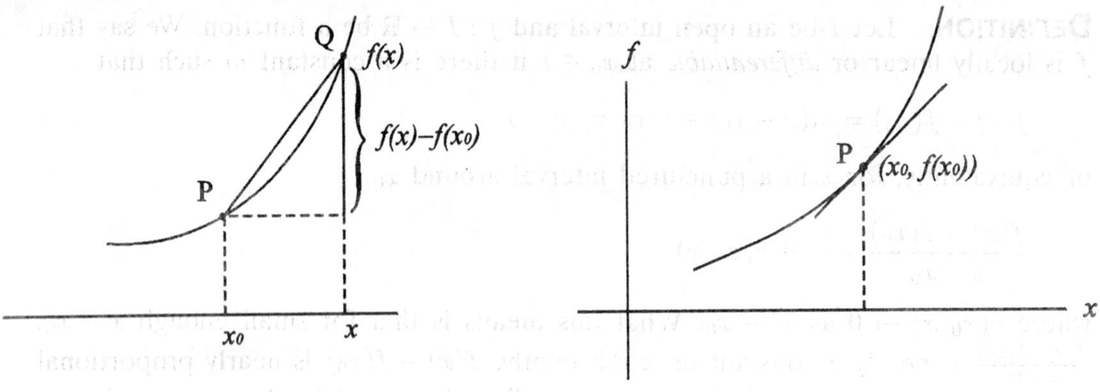

Figure 3.5

The chord PQ has the slope $\frac{f(x)-f(x_0)}{x-x_0}$. As x comes close to x_0, the chord tends to the tangent to the curve at $(x_0, f(x_0))$. So $\lim_{x\to x_0} \frac{f(x)-f(x_0)}{x-x_0}$ really represents the slope of the tangent at $(x_0, f(x_0))$ (see Figure 3.6).

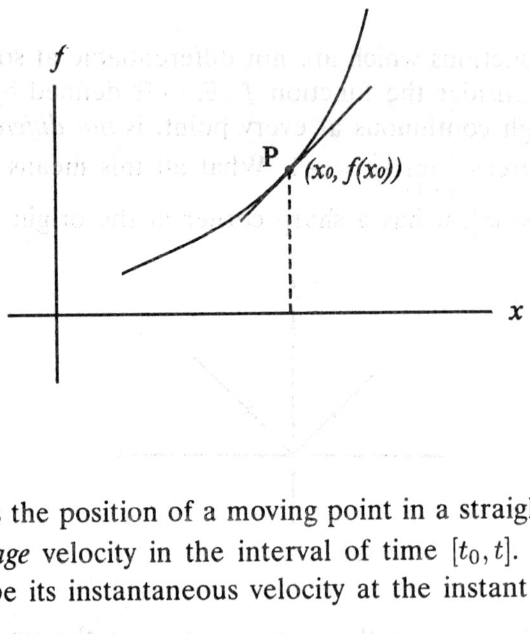

Figure 3.6

Similarly if $x(t)$ is the position of a moving point in a straight line at time t, then $\frac{x(t)-x(t_0)}{t-t_0}$ is its *average* velocity in the interval of time $[t_0, t]$. Its limit as t goes to t_0, if it exists, will be its instantaneous velocity at the instant of time t_0. We have

$$x'(t_0) = \lim_{t\to t_0} \frac{x(t)-x(t_0)}{t-t_0} = \text{instantaneous velocity at } t_0.$$

If the limit of $\frac{f(x)-f(x_0)}{x-x_0}$ does not exist as x tends to x_0, the curve $(x, f(x))$ cannot have a tangent at $(x_0, f(x_0))$, as we saw in the case of $f(x) = |x|$ at $(0,0)$; the graph abruptly changes its direction. If we look at the motion of a particle which is moving with uniform velocity till time t_0 and is abrupty brought to rest at that instant, then its graph would look as in Figure 3.7(a).

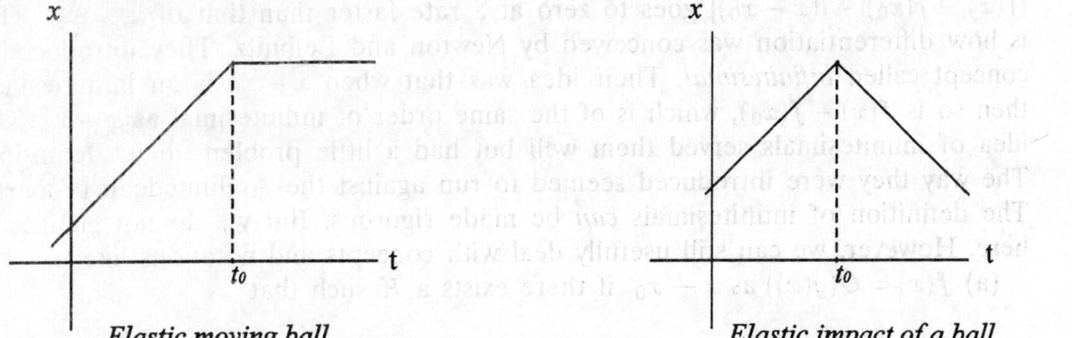

Figure 3.7

This is also what we think happens when a perfectly elastic ball impinges on another ball of the same mass at rest, or when a perfectly elastic ball moving at a constant speed impinges on a hard surface (see Figure 3.7(b)). We see that there is a sharp turn in the space-time graph of such a motion at time $t = t_0$. Recalling the interpretation of $x'(t) = \lim_{t \to t_0} \frac{x(t)-x(t_0)}{t-t_0}$ as its instantaneous velocity at $t = t_0$, we see that in the situation described above, instantaneous velocity at $t = t_0$ is not a meaningful concept.

We have already seen that continuous functions need not be differentiable at some points of their domain. Actually there are continuous functions which are not differentiable anywhere. This is something beyond the purview of this book. On the other hand, as the following result shows, every differentiable function is always continuous.

THEOREM: *If a function is differentiable at x_0 then it is continuous there.*

PROOF: If f is differentiable at x_0, then let $\lim_{x \to x_0} \frac{f(x)-f(x_0)}{x-x_0} = l$. Setting $r(x, x_0) = \frac{f(x)-f(x_0)}{x-x_0} - l$, we see that $\lim_{x \to x_0} r(x, x_0) = 0$. Thus we have

$$f(x) - f(x_0) = (x - x_0)l + (x - x_0)r(x, x_0).$$

Now

$$\lim_{x \to x_0} f(x) - f(x_0) = \lim_{x \to x_0} (x - x_0)l + \lim_{x \to x_0} (x - x_0)r(x, x_0) = 0.$$

This shows f is continuous at x_0. ■

Continuity of f at x_0 tells us that $f(x) - f(x_0)$ tends to zero as $x - x_0$ tends to zero. But in the case of differentiability, $f(x) - f(x_0)$ tends to zero at least as fast as $x - x_0$. The portion $l(x - x_0)$ goes to zero no doubt but the remainder $|f(x) - f(x_0) - l(x - x_0)|$ goes to zero at a rate faster than that of $|x - x_0|$. This is how differentiation was conceived by Newton and Leibnitz. They introduced a concept called *infinitesimal.* Their idea was that when $x - x_0$ is an infinitesimal, then so is $f(x) - f(x_0)$, which is of the same order of infinitesimal as $x - x_0$. The idea of infinitesimals served them well but had a little problem in its definition. The way they were introduced seemed to run against the Archimedean property. The definition of infinitesimals *can* be made rigorous. But we do not go into it here. However, we can still usefully deal with concepts and notations like:

(a) $f(x) = O(g(x))$ as $x \to x_0$ if there exists a K such that

$$|f(x)| \leq K|g(x)| \text{ for } x \text{ sufficiently near } x_0.$$

(b) $f(x) = o(g(x))$ as $x \to x_0$ if

$$\lim_{x \to x_0} \frac{f(x)}{g(x)} = 0.$$

Informally, $f(x) = o(g(x))$ means $f(x)$ is of smaller order than $g(x)$ as $x \to x_0$. In this notation, f is differentiable at x_0 if there is an l such that

$$|f(x) - f(x_0) - l(x - x_0)| = o(|x - x_0|).$$

We shall return to this point again. Let us first give examples of derivatives of some functions.

Examples

(The proofs of some statements will be given, the rest are left as exercises)

(a) $f(x) = x^n, f'(x_0) = \lim\limits_{x \to x_0} \dfrac{x^n - x_0^n}{x - x_0} = nx_0^{n-1}$, n a positive integer.

(b) $f(x) = x^n (x \neq 0,\ n$ a negative integer), $f'(x) = nx^{n-1}$.

(c) $f(x) = e^x,\ f'(x) = e^x$.

(d) $f(x) = a^x,\ f'(x) = a^x \log_e a$.

(e) $f(x) = \sin x$,

$$f'(x_0) = \lim_{x \to x_0} \frac{\sin x - \sin x_0}{x - x_0}$$

$$= \lim_{x \to x_0} \frac{2 \sin \dfrac{x - x_0}{2} \cos \dfrac{x + x_0}{2}}{x - x_0}$$

$$= \lim_{x \to x_0} \frac{\sin\left(\dfrac{x - x_0}{2}\right)}{\dfrac{x - x_0}{2}} \cos\left(\frac{x + x_0}{2}\right) = \cos x_0,$$

since $\cos x$ is continuous.

(f) $f(x) = \cos x$,

$$f'(x_0) = \lim_{x \to x_0} \frac{\cos x - \cos x_0}{x - x_0}$$

$$= \lim_{x \to x_0} \frac{2 \sin \dfrac{x_0 - x}{2} \sin\left(\dfrac{x_0 + x}{2}\right)}{x - x_0}$$

$$= \lim_{x \to x_0} - \frac{\sin \dfrac{x - x_0}{2}}{\dfrac{x - x_0}{2}} \cdot \sin\left(\frac{x + x_0}{2}\right) = - \sin x_0.$$

(g) $f(x) = \tan x$, $x \neq (2n+1)\frac{\pi}{2}$.

$$f'(x_0) = \lim_{x \to x_0} \frac{f(x) - f(x_0)}{x - x_0} = \lim_{x \to x_0} \frac{\dfrac{\sin x}{\cos x} - \dfrac{\sin x_0}{\cos x_0}}{x - x_0}$$

$$= \lim_{x \to x_0} \frac{\sin x \cos x_0 - \sin x_0 \cos x}{(x - x_0) \cos x \cos x_1}$$

$$= \lim_{x \to x_0} \frac{\sin(x - x_0)}{x - x_0} \cdot \frac{1}{\cos x \cos x_0} = \frac{1}{\cos^2 x_0} = \sec^2 x_0.$$

(h) $f(x) = \cot x \quad (x \neq n\pi)$, $f'(x_0) = - \operatorname{cosec}^2 x_0$.

(i) $f(x) = \sec x \quad (x \neq (2n+1)\frac{\pi}{2})$, $f'(x_0) = \sec x_0 \tan x_0$.

(j) $f(x) = \operatorname{cosec} x \quad (x \neq n\pi)$, $f'(x_0) = - \operatorname{cosec} x_0 \cot x_0$.

Differentiation of the Sum, Difference, Product and Quotient of Two Functions

THEOREM: *If $f, g : I \to \mathbb{R}$ are differentiable at $x_0 \in I$, where I is an open interval in $\mathbb{R}$, then so are $f \pm g$, fg at x_0. Furthermore, $\frac{f}{g}$ is differentiable at x_0 if $g(x_0) \neq 0$. We have*

(a) derivative of $f \pm g$ at x_0 is $f'(x_0) \pm g'(x_0)$;

(b) derivative of $f \cdot g$ at x_0 is $f'(x_0)g(x_0) + f(x_0)g'(x_0)$;

(c) derivative of $\dfrac{f}{g}$ at x_0 is $\dfrac{f'(x_0)g(x_0) - g'(x_0)f(x_0)}{(g(x_0))^2}$ if $g(x_0) \neq 0$.

The proofs are straightforward and therefore omitted.
We also have

THEOREM: *Let I and J be two intervals in $\mathbb{R}$ and $f : I \to J$, and $g : J \to \mathbb{R}$ be differentiable at $x_0 \in I$ and $f(x_0) \in J$ respectively. Then $h \equiv g \circ f : I \to \mathbb{R}$ is differentiable at x_0 and*

$$h'(x_0) = g'(f(x_0))f'(x_0).$$

$[h = g \circ f$ is defined as $h(x) = g(f(x))]$.

PROOF: Let us write $y = f(x)$ so that by the continuity of f at x_0, we have that as $x \to x_0$, $y \to y_0 = f(x_0)$. Since g is differentiable at y_0 we have

$$g(y) - g(y_0) = (g'(y_0) + r_1(y, y_0))\ (y - y_0),$$

where $r_1(y, y_0) \to 0$ as $y \to y_0$. Again since f is differentiable at x_0, we have

$$f(x) - f(x_0) = (f'(x_0) + r_2(x, x_0))(x - x_0),$$

where $r_2(x, x_0) \to 0$ as $x \to x_0$. Thus we have

$$\begin{aligned} g(f(x)) - g(f(x_0)) &= (g'(f(x_0) + r_1(y, y_0))(f(x) - f(x_0)) \\ &= (g'(f(x_0)) + r_1(y, y_0))(f'(x_0) + r_2(x, x_0))(x - x_0) \\ &= g'(f(x_0))f'(x_0)(x - x_0) + (x - x_0)r_3, \end{aligned}$$

where

$$r_3 = g'(f(x_0)r_2(x, x_0) + f'(x_0)r_1(y, y_0) + r_1(y, y_0)r_2(x, x_0).$$

Surely, $r_3 \to 0$ as $x \to x_0$ and hence

$$h'(x_0) = g'(f(x_0))f'(x_0).$$ ■

The above result is often called the *chain rule.*

Differential Notation of Leibnitz

For a differentiable function f, it we write $y = f(x)$ and $y + \Delta y = f(x + \Delta x)$, then we get

$$\lim_{\Delta x \to 0} \frac{\Delta y}{\Delta x} = \lim_{\Delta x \to 0} \frac{f(x + \Delta x) - f(x)}{\Delta x} = f'(x).$$

The expression

$$\lim_{\Delta x \to 0} \frac{\Delta y}{\Delta x} \text{ is often written as } \frac{dy}{dx}.$$

It is *not* true that $\frac{dy}{dx}$ is the quotient of limits of Δy and of Δx because both of them tend to zero. It should rather be thought of as an operation; $\frac{d}{dx}$ is the operation of differentiation operating on the variable y so that we have

$$\frac{d}{dx}(y) = \frac{dy}{dx}.$$

The operator $\frac{d}{dx}$ has the property $\frac{d}{dx}(y + u) = \frac{dy}{dx} + \frac{du}{dx}$, $\frac{d}{dx}(cy) = c\frac{dy}{dx}$, and for f and g, two differentiable functions with the domain of g containing the range of f, if we write $y = f(x)$ and $u = g(y)$, we have $u = g(f(x))$, then the chain rule gives $\frac{du}{dx} = \frac{du}{dy} \cdot \frac{dy}{dx}$. In the case when $y = f(x)$ so that $\frac{dy}{dx} = f'(x)$ we write, following the German mathematician Leibnitz,

$$dy = f'(x)dx,$$

and dy and dx are called the differentials of y and x respectively.

Remark: Let $f : \mathbb{R}^2 \to R$ be a function. We say that f is differentiable at $(x_0, y_0) \in \mathbb{R}^2$ if we can find numbers A, B depending on (x_0, y_0) only, so that

$$f(x, y) - f(x_0, y_0) = A(x - x_0) + B(y - y_0) + R(x, y, x_0, y_0)$$

such that

$$\lim \frac{R(x, y, x_0, y_0)}{\sqrt{[(x-x_0)^2 + (y-y_0)^2]}} = 0 \text{ as } (x, y) \to (x_0, y_0).$$

Observe that

$$A = \lim_{x \to x_0} \frac{f(x, y_0) - f(x_0, y_0)}{x - x_0} \text{ and } B = \lim_{y \to y_0} \frac{f(x_0, y) - (x_0, y_0)}{y - y_0}.$$

We call A and B the partial derivatives of f with respect to x and y respectively at (x_0, y_0), and we write

$$A = \frac{\partial f}{\partial x}(x_0, y_0), \quad B = \frac{\partial f}{\partial y}(x_0, y_0).$$

Sometimes we also write $\frac{\partial f}{\partial x}(x_0, y_0) = f_x(x_0, y_0)$ and $\frac{\partial f}{\partial y}(x_0, y_0) = f_y(x_0, y_0)$. Again as before, $\frac{\partial}{\partial x}$ and $\frac{\partial}{\partial y}$ may be thought of as operators which, operating on a function, give its partial derivatives with respect to x and y respectively.

Suppose that $\phi : \mathbb{R}^2 \to \mathbb{R}$ is differentiable, and that x and $y : \mathbb{R} \to \mathbb{R}$ are differentiable functions. Furthermore, let $u : \mathbb{R} \to \mathbb{R}$ be defined by

$$u(t) = \phi(x(t), y(t)).$$

It is not difficult to show that **(Exercise!)**

$$u'(t) = \frac{\partial \phi}{\partial x} x'(t) + \frac{\partial \phi}{\partial y} y'(t).$$

In Leibnitz's notation it reads

$$du = \frac{\partial u}{\partial x} dx + \frac{\partial u}{\partial y} dy.$$

As an application of this idea, consider the notion of equipotential surface in electrostatics. An electrically charged (infinite) cylindrical conductor is one such and because of the symmetry, it is enough to look at only a horizontal section of the cylinder which is a closed curve, i.e., the problem gets reduced to a 2-dimensional one. This equipotential curve is given by the equation $\phi(x, y) =$ constant, say c, where ϕ is the real-valued potential function. What is the electric field outside the curve? Using Leibnitz's notation as described above, we can write $d\phi = \frac{\partial \phi}{\partial x} dx + \frac{\partial \phi}{\partial y} dy$. Since ϕ is a constant, the relation above reads as

$$\frac{\partial \phi}{\partial x} x'(t) + \frac{\partial \phi}{\partial y} y'(t) = 0,$$

Clearly g is continuous on $[a.b]$ and differentiable in (a,b) and $g(a) = g(b) = 0$. Thus by Rolle's theorem, there is a $c \in (a,b)$ such that $g'(c) = 0$. But

$$\begin{aligned} g'(x) &= F'(x) + n\left(\frac{(b-x)^{n-1}}{(b-a)^n}\right)F(a) \\ &= -\frac{(b-x)^{n-1}}{(n-1)!}f^{(n)}(x) + \frac{n(b-x)^{n-1}}{(b-a)^n}F(a). \end{aligned}$$

$$\text{Thus}\quad 0 = g'(c) = -\frac{(b-c)^{n-1}}{(n-1)!}f^{(n)}(c) + n\frac{(b-c)^{n-1}}{(b-a)^n}F(a) \text{ or } F(a) = \frac{(b-a)^n}{n!}f^{(n)}(c),$$

which leads to

$$\begin{aligned} F(a) &= f(b) - f(a) - (b-a)f'(a) - \frac{(b-a)^2}{2!}f''(a) - \cdots \\ &\quad - \frac{(b-a)^{n-1}}{(n-1)!}f^{(n-1)}(a) \\ &= \frac{(b-a)^n}{n!}f^{(n)}(c) \end{aligned}$$

so that

$$f(b) = f(a) + (b-a)f'(a) + \cdots + \frac{(b-a)^{n-1}}{(n-1)!}f^{(n-1)}(a) + \frac{(b-a)^n}{n!}f^{(n)}(c). \qquad \blacksquare$$

Taylor's theorem gives an approximation of a sufficiently smooth function by polynomials. In fact if for some function f, which is arbitrarily often differentiable and $\frac{|(b-a)|^n|f^{(n)}(c)|}{n!} \to 0$ as $n \to \infty$, we can write

$$f(x) = f(a) + (x-a)f'(a) + \frac{(x-a)^2}{2!}f''(a) + \cdots + \frac{(x-a)^n}{n!}f^{(n)}(a) + \cdots$$

for $|x-a| \leq |b-a|$. This is called the Taylor series of the function f at a.

For example if one takes $f(x) = \sin x$, then $f^{(n)}(c) = \sin\left(\frac{n\pi}{2} + c\right)$, thereby giving us

$$\left|\frac{(b-a)^n f^{(n)}(c)}{n!}\right| \leq \frac{|b-a|^n}{n!}.$$

$\frac{f(c)-f(x)}{c-x} \geq 0$ if $c > x$, whereas $\frac{f(c)-f(x)}{c-x} \leq 0$ if $c < x$. We get by taking the left hand and right hand limits respectively $f'(c) \geq 0$ as well as $f'(c) \leq 0$ (as both left and right limits must exist in order that f is differentiable at c). This is possible only if $f'(c) = 0$. ∎

Geometrically, what Rolle's theorem says is that if a continuous smooth curve has equal ordinates at two different points, then there is at least one point on it where the tangent is parallel to the x-axis.

Algebraically, it means that between two roots of an equation $f(x) = 0$ there is a root of $f'(x) = 0$. Thus this theorem helps us to locate the roots of the equation $f'(x) = 0$ if we know already the roots of the equation $f(x) = 0$. The following result generalizes Rolle's theorem.

MEAN VALUE THEOREM: *If $f : [a, b] \to \mathbb{R}$ is continuous and $f'(x)$ exists for $a < x < b$, then there is a c such that $a < c < b$ and $f(b) - f(a) = (b - a)f'(c)$.*

Geometrically, this means that if the graph of the function f is smooth, then there is a point on the graph $(c, f(c))$, between the points $(a, f(a))$ and $(b, f(b))$, at which the tangent is parallel to the chord joining $(a, f(a))$ and $(b, f(b))$. One can then easily convince oneself that this is indeed a generalization of Rolle's theorem.

PROOF: Consider $F(x) = Af(x) + Bx$. We can choose $A = 1$ and $B = \frac{f(a)-f(b)}{b-a}$ so that $F(a) = F(b)$. Since f is continuous in the closed interval $[a, b]$ and differentiable in the open interval (a, b), F has the same property. Thus F satisfies all the hypotheses of Rolle's theorem and therefore there must exist a point c in the open interval (a, b) such that $F'(c) = 0$, or equivalently, $f'(c) = \{f(b) - f(a)\}/(b - a)$. ∎

Successive differentiation: We saw that if $x(t)$ is the position of a particle at time t, then its velocity at time t is given by $x'(t)$. Let $v(t) = x'(t)$, the velocity at the instant t, which is a function of time t. Then the acceleration of the particle being the rate of change of velocity would be, by the same token, $v'(t)$ (if it exists). We are thus led to consider the derivative of the derivative. If $f : (a, b) \to \mathbb{R}$ is differentiable at every $x \in (a, b)$, we write its derivative at each x as $f'(x) \in \mathbb{R}$. Thus we are led to a new function $f' : (a, b) \to \mathbb{R}$. If f' is differentiable, then we call its derivative the second derivative of f and write f'' for it. In Leibnitz's notation if $y = f(x)$ and $\frac{dy}{dx} = f'(x)$, we write $\frac{d^2y}{dx^2}$ for $f''(x)$. It is not hard to see what would be the meaning of the nth derivative or the nth differential coefficient of a function. If a function $f : (a, b) \to \mathbb{R}$ has derivatives up to the nth order we say that it is n times differentiable. Also it is clear that if a function is n times differentiable, then its derivatives up to order $n - 1$ must be continuous. If f is differentiable n times, then we write $f^{(n)}(x)$ or $\frac{d^ny}{dx^n}$ for its nth derivative, and if its nth derivative is continuous, then we say that the function f is n times *continuously differentiable*.

LEIBNITZ'S THEOREM (for successive derivative of product of two functions): *If $f, g : (a, b) \to \mathbb{R}$ are n times differentiable then $h : (a, b) \to \mathbb{R}$, defined by $h(x) = f(x)g(x)$, is also n times differentiable and*

$$h^{(n)}(x) = f^{(n)}(x)g(x) + \binom{n}{1} f^{(n-1)}(x)g'(x) + \cdots + \binom{n}{r} f^{(n-r)}(x)g^{(r)}(x) + \cdots + f(x)g^{(n)}(x),$$

where $\binom{n}{r}(0 \le r \le n)$ are the binomial coefficients.

This can easily be proved by induction on n using the product formula.

In Leibnitz's notation the above formula reads as (with $u = f(x), v = g(x)$)

$$\frac{d^n}{dx^n}(uv) = \frac{d^n u}{dx^n}v + \binom{n}{1}\frac{d^{n-1}u}{dx^{n-1}}\frac{dv}{dx} + \binom{n}{2}\frac{d^{n-2}u}{dx^{n-2}}\frac{d^2 v}{dx^2} + \cdots + \binom{n}{r}\frac{d^{n-r}u}{dx^{n-r}}\frac{d^r v}{dx^r} + \cdots + u\frac{d^n v}{dx^n}.$$

If the function f is more than once differentiable, then we would expect that a repeated application of the reasoning in the proof of the mean value theorem would yield a better insight into the behaviour of f near a point.

TAYLOR'S THEOREM: *Let I be an open interval in $\mathbb{R}$ and that $f : I \to \mathbb{R}$ be n times differentiable. For $a, b \in I$ there is $c \in (a, b)$ such that*

$$f(b) = f(a) + (b - a)f'(a) + \frac{(b-a)^2}{2!}f''(a) + \cdots + \frac{(b-a)^{n-1}}{(n-1)!}f^{(n-1)}(a) + \frac{(b-a)^n}{n!}f^{(n)}(c).$$

PROOF: The proof is very similar to that of the mean value theorem. Let

$$F(x) = f(b) - f(x) - (b - x)f'(x) - \frac{(b-x)^2}{2!}f''(x) - \cdots - \frac{(b-x)^{n-1}}{(n-1)!}f^{(n-1)}(x)$$

and $g(x) = F(x) - \left(\frac{b-x}{b-a}\right)^n F(a).$

i.e., the vector $(\frac{\partial\phi}{\partial x}, \frac{\partial\phi}{\partial y})$ is orthogonal to the tangent vector $(x'(t), y'(t))$.

Recalling that the electric field $\bar{E}$ at a point (x, y) outside has components $E_1 = -\frac{\partial\phi}{\partial x}$ and $E_2 = -\frac{\partial\phi}{\partial y}$, we conclude that the electric field $\bar{E}$ is along the (outward) normal. For example, for a right circular cylinder, the electric field is along the radial direction of every section.

We have seen that if f is differentiable at x_0, then

$$f(x) - f(x_0) - f'(x_0)(x - x_0)$$

is of smaller order than $(x - x_0)$ as $x \to x_0$. Writing $\Delta y = f(x) - f(x_0), \Delta x = (x - x_0)$, we get

$$\Delta y = f'(x_0)\Delta x$$

which is of smaller order than Δx. In other words, we are claiming that the increment in y is proportional to the increment in x when it is 'small', which is the principle of proportional parts. We have put the word small within quotation marks as it is a relative term. Let us look at it in an example.

Let $f(x) = \sqrt{x}, \quad x \geq 0$. Then $f'(x) = \frac{1}{2\sqrt{x}}$ for $x > 0$. So $f(16) = 4$. $f(17) = f(16) + f'(16).1 + r.1$, since $\Delta x = 1$. This gives us $\sqrt{17} = 4 + \frac{1}{8} + r = 4.125 + r$, whereas computation on a hand calculator will give $\sqrt{17} = 4.1231\cdots$. This shows that $|r| < .002$ and thus the differential approximates the increment in $f(x)$ correct at least to the second place of decimal.

Some Properties of Differentiable Functions

ROLLE'S THEOREM: *Suppose that* $f : [a, b] \to \mathbb{R}$ *is continuous and that* f *is differentiable on* (a, b). *Then if* $f(a) = f(b)$, *there is a point* $c \in (a, b)$ *for which*

$$f'(c) = 0.$$

PROOF: Since f is a continuous function on $[a, b]$, it will have a maximum and a minimum on $[a, b]$. Let c and d be two points where the function f attains its maximum and minimum respectively. If $f(c) = f(d)$ then f is a constant and hence its derivative is zero for all x in $[a, b]$ so that the theorem is trivially true. So suppose $f(c) \neq f(d)$. Then $c = a$ or b will imply $d \neq a$ and $d \neq b$. Similarly $d = a$ or b will imply $c \neq a$ and $c \neq b$. Hence at least one of c or d is different from a as well as b. Suppose c is different from a and b. (The rest of the proof is similar if d is different from a and b.)

As $f(c)$ is the maximum, we get $f(c) - f(x) \geq 0$ for $c > x$ or $c < x$. So

Since we know that for all $a, b \in \mathbb{R}$ $\lim_{n\to\infty} \frac{|b-a|^n}{n!} = 0$, we can write (by taking $a = 0$) for $x \in \mathbb{R}$,

$$\sin x = x - \frac{x^3}{3!} + \frac{x^5}{5!} - \frac{x^7}{7!} + \cdots + \frac{(-1)^n x^{2n+1}}{(2n+1)!} + \cdots .$$

Similarly,

$$\cos x = 1 - \frac{x^2}{2!} + \frac{x^4}{4!} - \cdots + \frac{(-1)^n x^{2n}}{(2n)!} + \cdots .$$

Both the above series converge for every value of x. One can similarly show that

$$e^x = 1 + x + \frac{x^2}{2!} + \frac{x^3}{3!} + \cdots + \frac{x^n}{n!} + \cdots, \text{ for all } x,$$

$$\log_e(1+x) = x - \frac{x^2}{2} + \frac{x^3}{3} - \frac{x^4}{4} + \cdots + (-1)^{n-1}\frac{x^n}{n} + \cdots, \text{ for } |x| < 1,$$

$$\tan^{-1} x = x - \frac{x^3}{3} + \frac{x^5}{5} - \frac{x^7}{7} + \cdots + (-1)^n \frac{x^{2n+1}}{2n+1} + \cdots, \quad \text{for } |x| < 1.$$

Warning: An arbitrarily often differentiable function need not admit a Taylor series expansion. This raises the question: when $f_1, \cdots, f_n, \cdots$ is a sequence of real-valued functions with domains in $\mathbb{R}$ what can one say about convergence of the series $\sum_{n=1}^{\infty} f_n(x)$?

It is obvious that convergence will depend on x. For example, consider the series $\sum_{n=1}^{\infty} \frac{\cos nx}{n}$. It is clear that for $x = 0$, the series diverges, whereas for $x = \frac{\pi}{2}$, the series becomes $-\frac{1}{2} + \frac{1}{4} - \frac{1}{6} + \frac{1}{8} - \frac{1}{10} + \cdots$, which is conditionally convergent. Similarly when $x = \pi$, the series becomes $-1 + \frac{1}{2} - \frac{1}{3} + \frac{1}{4} - \frac{1}{5} + \cdots$, which is again conditionally convergent. Thus the convergence of the above series depends on the value of x. Now let

$$s_n(x) = f_1(x) + f_2(x) + \cdots + f_n(x) \text{ for } x \in D \subset \mathbb{R}.$$

We would like to know for which values of x, $(s_n(x))_{n=1}^{\infty}$ converges as a sequence. We have already seen that while $\sum_{n=0}^{\infty} \frac{x^n}{n!}$ converges for every value of x, $\sum_{n=1}^{\infty} (-1)^{n-1} \frac{x^n}{n}$ converges only for $-1 < x \leq 1$. Suppose $f_1, \cdots, f_n, \cdots$ are functions with common domain of definition $D \subset \mathbb{R}$ such that $\sum_{n=1}^{\infty} f_n(x)$ converges for every $x \in D$. Set

$$f(x) = \sum_{n=1}^{\infty} f_n(x) \quad \text{for } x \in D.$$

Now it is clear that f too is a function with domain D. We would like to know the nature of the function f given some information about the nature of each of the functions f_n. For example if each f_n is continuous on D would f be continuous on D? We can ask a similar question about f when each f_n is differentiable and so on. Take, for example,

$$f_1(x) = \frac{1}{1+x}, f_2(x) = -\frac{x}{(1+x)(1+2x)}, \cdots, f_n(x) = -\frac{x}{(1+nx)(1+(n-1)x)}.$$

We observe that each of $f_1, f_2, \cdots, f_n, \cdots$ is continuous in $[0,1]$ and

$$\begin{aligned} s_n(x) &= f_1(x) + f_2(x) + \cdots + f_n(x) \\ &= \frac{1}{1+nx}. \end{aligned}$$

$(s_n(x))_{n=1}^{\infty}$ converges for every $x \in [0,1]$ but the limit function

$$f(x) = \lim_{n\to\infty} s_n(x) = \begin{cases} 1 & \text{for } x = 0, \\ 0 & \text{for } 0 < x \leq 1, \end{cases}$$

which shows that f is discontinuous at 0. This should convince us that the continuity of each f_n in D and the convergence of $\sum f_n(x)$ for all $x \in D$ are not enough to infer that the sum function is continuous. We can construct even nastier examples where the sum function behaves quite at variance with each term of the series. So we introduce the following notion:

DEFINITION: Let $(f_n)_{n=1}^{\infty}$ be a sequence of functions with common domain D. Let $\lim_{n\to\infty} f_n(x) = f(x)$ for every $x \in D$. We say that the convergence is *uniform* if for all $\varepsilon > 0$ we can find an n_0 depending on ε *alone* but independent of $x \in D$ such that

$$|f_n(x) - f(x)| < \varepsilon \quad \text{for all } n > n_0 \quad \text{and for all } x \in D.$$

It can be shown that if $(f_n)_{n=1}^{\infty}$ is a sequence of function on D, then $(f_n(x))_{n=1}^{\infty}$ converges uniformly in D if and only if for all $\varepsilon > 0$ there exists an n_0 depending on ε but not on x such that

$$|f_m(x) - f_n(x)| < \varepsilon \text{ for all } m, n > n_0 \text{ for all } x \in D.$$

THEOREM: *Let $(f_n)_{n=1}^{\infty}$ be a sequence of continuous functions in $[a, b]$ converging uniformly to f. Then f is continuous in $[a, b]$.*

PROOF: Since $(f_n)_{n=1}^{\infty}$ is uniformly convergent, for all $\varepsilon > 0$ there exists an n_0 such that

$$|f_n(x) - f(x)| < \varepsilon/3 \text{ for all } n > n_0 \text{ and for all } x \in [a, b].$$

Fix an $n_1 > n_0$, and note that since f_{n_1} is continuous at any $x_0 \in [a, b]$, we can find a $\delta > 0$ such that

$$|f_{n_1}(x) - f_{n_1}(x_0)| < \varepsilon/3 \quad \text{for} \quad |x - x_0| < \delta.$$

$$\begin{aligned} \text{Thus } |f(x) - f(x_0)| &= |f(x) - f_{n_1}(x) + f_{n_1}(x) - f_{n_1}(x_0) + f_{n_1}(x_0) - f(x_0)| \\ &\leq |f(x) - f_{n_1}(x)| + |f_{n_1}(x) - f_{n_1}(x_0)| + |f_{n_1}(x_0) - f(x_0)| \\ &< \varepsilon/3 + \varepsilon/3 + \varepsilon/3 \quad \text{for } |x - x_0| < \delta. \end{aligned}$$

■

Remark: We remarked earlier that if we want to allow taking square roots of negative real numbers, we have to extend the field of real numbers to a larger field called complex numbers. One of the ways is to consider the set

$$\mathbb{C} = \{(x, y) : x, y \in \mathbb{R}\},$$

with addition and multiplication defined by

$$\begin{aligned} (x_1, y_1) + (x_2, y_2) &= (x_1 + x_2, y_1 + y_2) \\ (x_1, y_1)(x_2, y_2) &= (x_1x_2 - y_1y_2, x_1y_2 + x_2y_1) \end{aligned}$$

one easily sees that these are enough to make $\mathbb{C}$ into a field with $(0, 0)$ as the additive identity, $(1, 0)$ as the multiplicative identity. If $z = (x, y)$, then $-z$, the additive inverse of z, shall be $(-x, -y)$. We see that

$$\{(x, 0) : x \in \mathbb{R}\}$$

is a field with the same operations. Indeed the element $(x, 0)$ in $\mathbb{C}$ behaves as if it were the element $x \in \mathbb{R}$. So we may identify the subset $\{(x, 0) : x \in \mathbb{R}\}$ with $\mathbb{R}$. Thus $\mathbb{C}$ is deemed as an extension of $\mathbb{R}$.

We have observed that $(1, 0)$ can be identified with 1 of $\mathbb{R}$. If we now multiply $(0, 1)$ by $(0, 1)$ we have

$$(0, 1)(0, 1) = (-1, 0) = -(1, 0).$$

If we now write $(0,1)$ as i and $(1,0)$ as 1 then we have

$$i^2 = -1.$$

This makes taking square roots of negative numbers possible in $\mathbb{C}$. Further, for a real number α and $z = (x, y)$ if we write

$$\alpha z = (\alpha x, \alpha y),$$

then it is easy to see that

$$z = (x, y) = x(1,0) + y(0,1) = x + yi$$

if we agree to write $x(1,0)$ as x and $(0,1)$ as i. It is the same thing as writing $\mathbb{C} = \{x + yi : x, y \in \mathbb{R}\}$. Such numbers as $x + yi$ are called *complex numbers*; they are added, subtracted and multiplied as if they were real numbers with the understanding that we replace i^2 by -1. So

$$\begin{aligned}(x_1 + y_1 i)(x_2 + y_2 i) &= x_1x_2 + x_1y_2 i + x_2y_1 i + y_1y_2 i^2 \\ &= (x_1x_2 - y_1y_2) + (x_1y_2 + x_2y_1)i.\end{aligned}$$

We define, for $x = x + yi$, the absolute value of z or the modules of z, written $|z|$, by

$$|z| = \sqrt{x^2 + y^2}.$$

Exercises

Verify

(i) $|z_1 z_2| = |z_1|\,|z_2| \;\; \forall \;\; z_1, z_2 \in \mathbb{C}$,

(ii) $|z_1 + z_2| \leq |z_1| + |z_2| \;\; \forall \;\; z_1, z_2 \in \mathbb{C}$.

With this we can define convergence of complex sequences and series the same way we did for reals. We can even define continuity of function $f : \mathbb{R} \to \mathbb{C}$ the same way. We easily see that for $z \in \mathbb{C}$ the series $1 + z + \frac{z^2}{z!} + \cdots + \frac{z^n}{n!} + \cdots$ converges and we define, in analogy with the series expansion of e^x by the series for $x \in \mathbb{R}$,

$$e^z = 1 + z + \frac{z^2}{z!} + \cdots + \frac{z^n}{n!} \cdots.$$

For $\theta \in \mathbb{R}$ we see that

$$
\begin{aligned}
e^{i\theta} &= 1 + i\theta + \frac{(i\theta)^2}{z!} + \frac{(i\theta)^3}{3!} + \cdots \\
&= \left(1 - \frac{\theta^2}{2!} + \frac{\theta^4}{4!} - \cdots\right) + i\left(\theta - \frac{\theta^3}{3!} + \cdots\right) \\
&= \cos\theta + i\sin\theta.
\end{aligned}
$$

This is the famous Euler's formula.

3.3 Integration

Historically, the problem of integration has its origin in the problems of finding areas under a given curve and volumes of solids of revolution. Consider the problem of finding the area of the region $\Omega = \{(x, y) : 0 \le x \le 1, 0 \le y \le x^2\}$ (Figure 3.8)

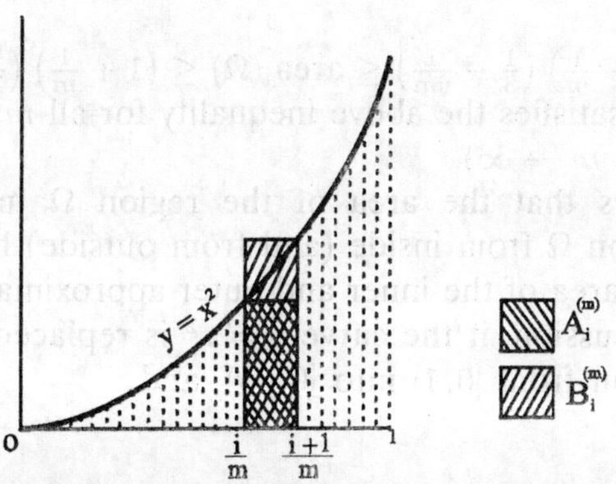

Figure 3.8

bounded by the x-axis, the line x=1 and the curve $y = x^2$. Fix $m \ge 1$, and for $i = 0, 1, \cdots, m-1$, let

$$
\begin{aligned}
A_i^{(m)} &= \left\{(x, y) : \frac{i}{m} \le x < \frac{i+1}{m},\quad 0 \le y \le \frac{i^2}{m^2}\right\} \\
B_i^{(m)} &= \left\{(x, y) : \frac{i}{m} \le x < \frac{i+1}{m},\quad 0 \le y \le \frac{(i+1)^2}{m^2}\right\}
\end{aligned}
$$

Also set $A^{(m)} = \bigcup_{i=0}^{m-1} A_i^{(m)}$, $B^{(m)} = \bigcup_{i=0}^{m-1} B_i^{(m)}$. It is easy to see that

$$A^{(m)} \subseteq \Omega \subseteq B^{(m)},$$

and thus $\text{area}(A^{(m)}) \leq \text{area}(\Omega) \leq \text{area}(B^{(m)})$.

Now $A^{(m)}$ and $B^{(m)}$ are (disjoint) unions of rectangles $A_i^{(m)}$ and $B_i^{(m)}$ respectively, which means that

$$\begin{aligned}\text{area}(A^{(m)}) &= \sum_{i=0}^{m-1} \text{area}(A_i^{(m)}) = \sum_{i=0}^{m-1} \frac{1}{m} \cdot \frac{i^2}{m^2} \\ &= \frac{1}{m^3} \frac{(m-1)m(2m-1)}{6} = \left(1 - \frac{1}{m}\right)\left(\frac{1}{3} - \frac{1}{6m}\right)\end{aligned}$$

and similarly

$$\text{area}(B^{(m)}) = \sum_{i=0}^{m-1} \frac{1}{m} \frac{(i+1)^2}{m^2} = \left(1 + \frac{1}{m}\right)\left(\frac{1}{3} + \frac{1}{6m}\right).$$

Thus we have that $\left(1 - \frac{1}{m}\right)\left(\frac{1}{3} - \frac{1}{6m}\right) \leq \text{area}\ (\Omega) \leq \left(1 + \frac{1}{m}\right)\left(\frac{1}{3} + \frac{1}{6m}\right)$ for all m. The only real number that satisfies the above inequality for all m is $1/3$ (try to establish this by taking limits as $m \to \infty$).

Therefore, it follows that the area of the region Ω must be $1/3$. Here we approximated the region Ω from inside (and from outside) by unions of rectangles in such a way that the area of the inner and outer approximations converged to the same limit. In this discussion, if the curve $y = x^2$ is replaced by $y = f(x)$, where f is a continuous function from $[0, 1]$ into $[0, \infty)$, and

$$\begin{aligned}A_i^{(m)} &= \left\{(x, y) : \frac{i}{m} \leq x < \frac{i+1}{m}, 0 \leq y \leq \alpha_i^{(m)}\right\}, \\ B_i^{(m)} &= \left\{(x, y) : \frac{i}{m} \leq x < \frac{i+1}{m}, 0 \leq y \leq \beta_i^{(m)}\right\}, \\ \text{where } \alpha_i^{(m)} &= \inf\left\{f(x) : \frac{i}{m} \leq x < \frac{i+1}{m}\right\}, \\ \beta_i^{(m)} &= \sup\left\{f(x) : \frac{i}{m} \leq x < \frac{i+1}{m}\right\},\end{aligned}$$

then as before, it is clear that

$$A^{(m)} = \bigcup_{i=0}^{m-1} A_i^m \subseteq \Omega \subseteq B^{(m)} = \bigcup_{i=0}^{m-1} B_i^{(m)}.$$

Here again, it can be shown (using uniform continuity of f) that area $(A^{(m)})$ and area $(B^{(m)})$ converge to the same number, which must then be the area of the region Ω. We will prove this assertion later, but readers are encouraged to prove it themselves.

The area of the region Ω is called the integral of $f(x)$ over the interval $[0,1]$. If the function f is not continuous, it is not clear if the outer and inner approximations converge to the same limit!

In the light of the above discussion, we are led to the following definitions:

DEFINITION: By a partition P of a given interval $[a, b]$, we mean a finite set $P = \{x_0, x_1, \cdots, x_n\}$ of points in $[a, b]$ such that $a = x_0 < x_1 < x_2 < \cdots < x_n = b$. Let $f : [a, b] \to \mathbb{R}$ be a bounded function, $P = \{a = x_0 < x_1 < \cdots < x_n = b\}$ be a partition of [a,b] and let

$$U(P, f) = \sum_{i=1}^{n} \left\{ \sup_{x_{i-1} \le x \le x_i} f(x) \right\} (x_i - x_{i-1})$$

$$L(P, f) = \sum_{i=1}^{n} \left\{ \inf_{x_{i-1} \le x \le x_i} f(x) \right\} (x_i - x_{i-1}).$$

It is clear that

$$(b-a) \inf_{x \in [a,b]} f(x) \le L(P, f) \le U(P, f) \le (b-a) \sup_{x \in [a,b]} f(x).$$

Define

$$U\text{-}\int_a^b f(x)dx = \inf U(P, f),$$

$$L\text{-}\int_a^b f(x)dx = \sup L(P, f),$$

where the infimum and supremum are taken over all partitions P of $[a, b]$.

$U\text{-}\int_a^b f(x)\,dx$ and $L\text{-}\int_a^b f(x)\,dx$ are called the upper integral and the lower integral respectively of f over $[a,b]$.

DEFINITION: A bounded function $f:[a,b]\to\mathbb{R}$ is said to be *Riemann integrable* on $[a,b]$ if the upper and lower integrals are equal, and then the common value is called the (Riemann) *integral* of f and is denoted by $\int_a^b f(x)dx$.

By taking the partition $P=\{a,b\}$, it follows that if f is Riemann integrable, then

$$\left(\inf_{a\le x\le b} f(x)\right)(b-a)\le\int_a^b f(x)dx\le\left(\sup_{a\le x\le b} f(x)\right)(b-a).$$

For partitions $P_1=\{a=x_0<x_1<\cdots<x_n=b\}$ and $P_2=\{a=y_0<y_1<\cdots<y_m=b\}$, we say that P_2 is a refinement of P_1 if $\{x_0,x_1,\cdots,x_n\}\subseteq\{y_0,y_1,\cdots,y_m\}$.

We leave it to the reader to verify the following properties of $L(f,P)$ and $U(f,P)$:

1. If P_2 is a refinement of P_1, then
$$L(f,P_1)\le L(f,P_2)\le U(f,P_2)\le U(f,P_1).$$

2. Given two partitions P,P' of $[a,b]$, there exists a partition P'' that is a refinement of P as well of P' (in fact, P'' is obtained by listing all the points of P and P' in order).

3. For any partition P of $[a,b]$,
$$L(f,P)\le L\text{-}\int_a^b f(x)dx\le U\text{-}\int_a^b f(x)dx\le U(f,P).$$

4. For any partition P and bounded functions f_1,f_2,
$$L(f_1+f_2,P)\ \ge\ L(f_1,P)+L(f_2,P),$$
$$U(f_1+f_2,P)\ \le\ U(f_1,P)+U(f_2,P).$$

The following is an important result on integration and useful for computing integrals.

THEOREM: *Let f be a bounded function. Suppose there exists a sequence P_n of partitions of $[a,b]$ such that $\lim_n L(f,P_n)=\lim_n U(f,P_n)=\alpha$ (say), then f is Riemann integrable and $\alpha=\int_a^b f(x)dx$.*

PROOF: By Property 3 mentioned above, $\alpha \leq L\text{-}\int_a^b f(x)dx \leq U\text{-}\int_a^b f(x)dx \leq \alpha$ and thus the result follows. ■

Thus, the computation at the beginning of the section shows that $f(x) = x^2$ is Riemann integrable over [0,1] and that $\int_0^1 f(x)dx = \frac{1}{3}$.

Here is another useful result.

THEOREM: *Let f be Riemann integrable and $\varepsilon > 0$ be arbitrary. Then there exists a partition P such that $U(f,P) - L(f,P) < \varepsilon$.*

PROOF: Let P_1, P_2 be partitions such that $|U(f,P_1) - \int_a^b f(x)dx| < \varepsilon/2$ and $|L(f,P_2) - \int_a^b f(x)dx| < \varepsilon/2$.

Let P be a common refinement of P_1 and P_2 (see Property 2 above). Then by Properties 1 and 3 above

$$\int_a^b f(x)dx \leq U(f,P) \leq U(f,P_1)$$

and $\displaystyle\int_a^b f(x)dx \geq L(f,P) \geq L(f,P_2)$

and hence

$$|U(f,P) - \int_a^b f(x)dx| < \varepsilon/2,$$

$$|L(f,P) - \int_a^b f(x)dx| < \varepsilon/2$$

from which the result follows. ■

We shall now prove that every continuous function on $[a,b]$ is Riemann integrable.

THEOREM: *Let f be a continuous function on $[a,b]$. Then f is Riemann integrable.*

PROOF: For $n \geq 1$, define $P^{(n)} = \{a = x_0 < \cdots < x_{2^n} = b\}$, where $x_i = a + \frac{i}{2^n}(b-a)$. Clearly $P^{(n+1)}$ is a refinement of $P^{(n)}$. Thus $L(f,P^{(n)}) \leq L(f,P^{(n+1)}) \leq U(f,P^{(n+1)}) \leq U(f,P^{(n)})$. Let α be the limit of the (increasing) sequence $L(f,P^{(n)})$ and β be the limit of the decreasing sequence $U(f,P^{(n)})$ (these limits exist because the sequences are monotone and bounded). Clearly we have that

$$L(f,P^{(n)}) \leq \alpha \leq L\text{-}\int_a^b f(x)dx \leq U\text{-}\int_a^b f(x)dx \leq \beta \leq U(f,P^{(n)}).$$

To complete the proof, we will show that $\alpha = \beta$. Fix $\varepsilon > 0$. Using uniform continuity of the continuous function f on $[a, b]$, choose $\delta > 0$ such that $|f(x) - f(y)| < \varepsilon$ whenever $|x - y| < \delta$. Next, choose n such that $\frac{(b-a)}{2^n} < \delta$. Taking $x_i = a + \frac{i}{2^n}(b - a)$ and $P^{(n)} = \{x_0, x_1, \dots, x_n\}$ so that one has $|\sup_{x_{i-1} \le x \le x_i} f(x) - \inf_{x_{i-1} \le x \le x_i} f(x)| < \varepsilon$ for $1 \le i \le n$ and hence $|U(f, P^{(n)}) - L(f, P^{(n)})| \le \varepsilon(b - a)$. By taking the limit as $n \to \infty$ this leads to $|\beta - \alpha| \le \varepsilon(b - a)$, and since this holds for every ε, it follows that $\alpha = \beta$. ■

An important property of the Riemann integral is its linearity: if f_1 and f_2 are Riemann integrable, then $f = (f_1 + f_2)$ is also Riemann integrable and

$$\int_a^b (f_1 + f_2)(x)dx = \int_a^b f_1(x)dx + \int_a^b f_2(x)dx.$$

This follows from the Property 4 of $U(f, P), L(f, P)$ listed above.

Given a function f on $[a, b]$ which is Riemann integrable on $[a, b]$, it is easy to see that it is also Riemann integrable on any subinterval of $[a, b]$ and further, for any $c, a < c < b$ one has

$$\int_a^b f(x)dx = \int_a^c f(x)dx + \int_c^b f(x)dx.$$

We will now give two important results that connect integration and differentiation. Also, these will relate the Riemann integral with indefinite integrals that most readers must have computed earlier and with which they are familiar.

THEOREM: *Suppose f is Riemann integrable on $[a, b]$ and for $a \le c \le b$ define $F(c) = \int_a^c f(x)dx$. Then F is continuous. If, furthermore, f is continuous at t, $a < t < b$, then F is differentiable at t and*

$$F'(t) = f(t).$$

PROOF: Continuity of F is easy to see. For $h > 0$ such that $t + h \le b$, we have

$$\begin{aligned} \frac{F(t+h) - F(t)}{h} - f(t) &= \frac{1}{h}\int_t^{t+h} f(x)dx - f(t) \\ &= \frac{1}{h}\int_t^{t+h} (f(x) - f(t))dx. \end{aligned}$$

Now given $\varepsilon > 0$, choose $\delta > 0$ such that $|f(x) - f(t)| < \varepsilon/2$ whenever $|x - t| < \delta$. Then for $h < \delta$, one has $|f(x) - f(t)| < \varepsilon$ for $t \leq x \leq t + h$, and hence

$$-\varepsilon h \leq \int_t^{t+h} (f(x) - f(t))dx \leq \varepsilon h.$$

Thus, for $0 < h < \delta$,

$$\left| \frac{F(t+h) - F(t)}{h} - f(t) \right| < \varepsilon.$$

We can similarly show that this is true for $-\delta < h \leq 0$. This shows that F is differentiable and $F'(t) = f(t)$. ■

THEOREM (FUNDAMENTAL THEOREM OF INTEGRAL CALCULUS): *Let f be a Riemann integrable function on $[a, b]$ and F be a continuous function on $[a, b]$ that is differentiable on (a, b) and*

$$F'(x) = f(x), \; a < x < b.$$

Then

$$\int_a^b f(x)dx = F(b) - F(a).$$

PROOF: Let P be any partition of $[a, b]$, say $P = \{x_0, x_1, \cdots, x_m\}$ with $a = x_0 < x_1 < x_2 < \cdots < x_m = b$. Fix i, $0 \leq i \leq m - 1$. Using the mean value theorem for the function F on $[x_i, x_{i+1}]$, we get that there exists $c_i \in (x_i, x_{i+1})$ such that

$$F(x_{i+1}) - F(x_i) = f(c_i)(x_{i+1} - x_i).$$

Thus

$$\left[\inf_{x_i \leq x \leq x_{i+1}} f(x) \right] (x_{i+1} - x_i) \leq F(x_{i+1}) - F(x_i) \leq \left[\sup_{x_i \leq x \leq x_{i+1}} f(x) \right] (x_{i+1} - x_i).$$

Summing from $i = 0$ to $m-1$ and noting that $\sum_{i=0}^{m-1} [F(x_{i+1}) - F(x_i)] = F(b) - F(a)$, we have

$$L(f, P) \leq F(b) - F(a) \leq U(f, P).$$

This implies that $L\text{-}\int_a^b f(x)dx \leq [F(b) - F(a)] \leq U\text{-}\int_a^b f(x)dx$. Since f is Riemann integrable, the upper and lower integrals are equal and this shows that

$$\int_a^b f(x)dx = F(b) - F(a).$$

■

In view of the previous result, any function F such that $F' = f$ is called an indefinite integral of f (written as $\int f = F$). The (Riemann) integral $\int_a^b f dx$ is also called the definite integral of f over $[a, b]$.

You must have come across many techniques to find indefinite integrals. A very useful result for the computation of indefinite integrals is the integration by parts formula which we will prove.

THEOREM: *Let f, g be continuous functions on $[a, b]$. Let $F = \int f dx, G = \int g dx$ be indefinite integrals of f, g respectively. Then*

$$\int_a^b F(x)g(x)dx = F(b)G(b) - F(a)G(a) - \int_a^b f(x)G(x)dx.$$

PROOF: Let $H(x) = F(x)G(x)$. Since $F' = f$ and $G' = g$, it follows that

$$H' = Fg + fG.$$

Thus $$H(b) - H(a) = \int_a^b F(x)g(x)dx + \int_a^b f(x)G(x)dx.$$

This yields the required result. ■

Suggested Reading

1. *Calculus*, Vol. I by T.M. Apostol, Wiley Eastern, New Delhi 1980 (read Chapters 1, 3, 4 and 5).

2. *Calculus with Applications and Computing* by P. Lax, S. Burstein and A. Lax, Narosa Publishing House, New Delhi 1979 (read Chapters 2, 3 and 4).

3. *Calculus: An intuitive and Physical Approach* by M. Kline, John Wiley 1967.

Chapter 4

Differential Equations and other Applications of Calculus

"In addition to the straight lines, circles, planes and spheres with which every student of Euclid is familiar, the Greeks knew the properties of the curves given by cutting a cone with a plane – the ellipse, parabola and hyperbola. Kepler discovered by analysis of astronomical observations, and Newton proved mathematically on the basis of the inverse square law of gravitational attraction, that the planets move in ellipses. The geometry of ancient Greece thus became the cornerstone of modern astronomy."

J.L. Synge

4.1 Anti-derivatives

By the fundamental theorem of integral calculus, we can construct the function f given f'; in such a case f is called the anti-derivative of f'. In fact, if we are given that $f'(x) = g(x)$ for all $x \in [a, b]$, then we can write $f(x) = \int_a^x g(t)dt + C$, where C is the constant of integration and can be determined explicitly if we know the value of f at some point c in the interval $[a, b]$ by setting $C = f(c) - \int_a^c g(t)dt$.

One can also visualize the above construction graphically by recalling that the slope of f at x is given by $f'(x)$. Thus in an interval where $f' \geq 0$, f is non-decreasing and where $f' \leq 0, f$ is non-increasing. In Figure 4.1 we have sketched two approximate graphs of f on the right, given that of f' on the left. Note that the constants of integration c in the two cases are determined by $f(0) = 0$ and $f(0) = 1$ respectively.

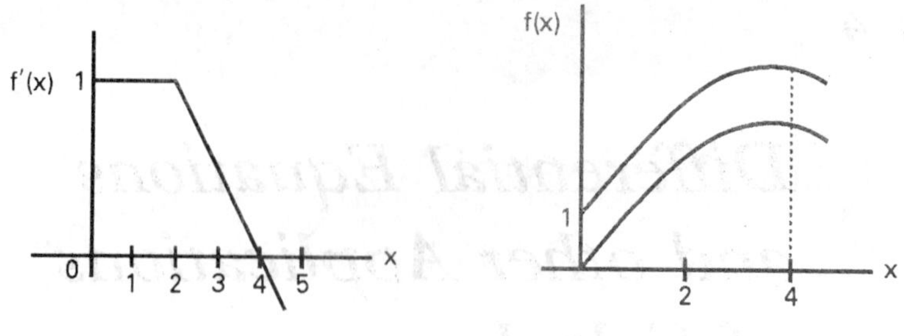

Figure 4.1

In chapter 3, we talked about velocity being the derivative of position (looked upon as a function of time) and acceleration as the derivative of velocity with respect to time. As in the above, we can try to go 'backward' from acceleration to velocity to position. At first we shall consider the case of *uniform motion*, e.g., consider a car moving in a straight line with a constant velocity of 40 kmph (kilometres per hour). Then if s is the distance (in km) from a fixed point on the line and if t is time (in hours), then we can describe the motion by stating that the velocity, $\frac{ds}{dt}$, equals 40, or

$$\frac{ds}{dt} = 40.$$

Such an equation is a simple example of a *differential equation* and we can solve it by the method of anti-derivatives as described above to get $s = \int_0^t (40)dx + s_0 = 40t + s_0$, where s_0 is the distance of the initial point (at $t = 0$) from the reference point.

The next interesting model is that of *uniformly accelerated motion*, particularly the motion of a body under the influence of gravity. It has been known since Galileo's time that bodies 'close' to earth experience an attraction leading to a constant acceleration g, directed towards the centre of the earth. The experimental value of g has been found to be 9.8 m/sec^2 (metre per second per second). Let s be the distance measured upwards from the surface of the earth and v, the upward velocity. Then the above can be expressed by the differential equation

$$\frac{dv}{dt} = \frac{d^2s}{dt^2} = -g.$$

This can be solved in two stages; first, we find v by the method of anti-derivatives to get $v = -gt + v_0$ (v_0 is the velocity at time $t = 0$); next we find s by applying the same method to $\frac{ds}{dt} = v = -gt + v_0$ to get $s = -\frac{1}{2}gt^2 + v_0t + s_0$. Eliminating t from the above two expressions, we get the third familiar form of the equation of motion, viz., $v_0^2 - v^2 = 2g(s - s_0)$.

4.2 Differential Equations

One can draw the following conclusions from the above discussions:

(i) In nature, for many aspects of human activities and in their interplay, there are always a few primitive objects (like distance and time as in the previous paragraph; population of fish in a pond; supply of a consumable commodity and its price, etc.) which are measurable; (ii) these measurable objects are usually changing with respect to each other and one experimentally or otherwise establishes or postulates a kind of relationship between these rates of changes and each other. Once one has such a description, we say that we have a model and in special cases, where this description is backed by experiment, we say we have a law (as in the case of Galileo's law of falling bodies, Newton's laws, etc.); (iii) the law or the model often sets up an equation relating rates of change and various observable quantities, in many cases leading to what we have called differential equations; (iv) finally, one solves the equations leading to the sought after relations between the observable quantities. This kind of modelling of various phenomena and solving them to compare with observation is one of the most successful applications of the subject of mathematical analysis. Next we turn to the formal definitions of a differential equation and its solution.

Let y be a function of $x \in [a, b] \subseteq \mathbb{R}$, and denote by $y', y'', y''', \cdots, y^{(n)}$ the derivatives of y with respect to x of order 1, 2, 3, $\cdots$, n respectively. An (ordinary) *differential equation of nth order* is a (functional) relation between $x, y', y'', y''', \cdots, y^{(n)}$, viz., $F(x, y, y', y'', y''', \cdots, y^{(n)}) = 0$, where F is a given function of $(n + 2)$ real variables. Thus in the two examples relating to motion that we have seen so far, the first one was of the first-order while the second one, of order 2. There was something still more special in these two examples: in fact the dependences on s, s', s'', etc., were linear. More generally, a differential equation of order n is linear if the defining function F is of the form $F = \sum_{j=0}^{n} a_j(x)y^{(j)}$, where we have set $y^{(0)} = y$. Here we shall content ourselves with studying first- and second-order equations only, and that too mostly linear ones.

Given a differential equation described by F as above, what do we mean by its solution? To solve such an equation, we need to find one or more n-times differentiable functions $y = y(x)$ such that it, along with its derivatives upto and including the nth order, satisfies the defining equation.

As we had said earlier, we have so far talked about ordinary differential equations. There are many natural phenomena in which the observable quantity is a function of more than one independent variable. Consider, for example, the motion of waves in a lake after you have thrown a small pebble into it. If one imagines oneself to be sitting on a water particle, then one will experience a periodic up and down movement with time, while if one observes the surface of the lake from outside, one can see that the same state of displacement appears periodically in space at any instant of time. Thus the observable quantity, displacement of the water particle from its equilibrium before any stone was thrown, is a function of space, say x and y (the two variables one needs to describe the points on the surface of the lake), and time t. One can also model this or similar phenomena by relating various derivatives of the unknown function with respect to the different independent variables (the so-called partial derivatives) with each other, leading to what is called a partial differential equation, and can study its solution. However, the study of such equations is well beyond the scope of this book.

4.3 First-order differential equation

We have already come across one such equation, namely the first example in Section 4.1, which was also linear. More generally, a first-order linear differential equation will have the form $a_0(x)y' + a_1(x)y = g(x)$, where a_0, a_1 and g are given continuous functions on an interval [a,b], and a_0 does not vanish anywhere in the interval. In such a case we rewrite the equation by dividing by a_0 to get

$$\frac{dy}{dx} + P(x)y = Q(x).$$

If Q = 0, then the equation is said to be *homogeneous*, otherwise, *inhomogeneous*. Such an equation is solved by the method of *integrating factor*. For this, we first solve the homogeneous equation, viz., $y' = -P(x)y$ or $y^{-1}y' = -P(x)$. Recalling that by the chain rule $\frac{d}{dx}(\log y) = y^{-1}\frac{dy}{dx}$, we can rewrite this as $\frac{d}{dx}(\log y) = -P(x)$, we are ready to apply the fundamental theorem of integral calculus (or the method of anti-derivatives as we had christened it in Section 4.1). This leads to

$$\log y = -\int_a^x P(z)dz + C_1,$$

or $y \equiv y_h = Ce^{-\int_a^x P(z)dz}$, where we have written $C = e^{C_1}$. Next we want to solve the full inhomogeneous equation. For this we multiply both sides of the equation by the function $\exp(\int_a^x P(z)dz)$ (which, we note, is the inverse of the homogeneous solution upto a constant and is càlled the *integrating factor*) and observe that the resulting equation can be rewritten as

$$e^{\int_a^x P(z)dz}\left(\frac{dy}{dx} + P(x)y\right) = Q(x)e^{\int_a^x P(z)dz},$$

or

$$\frac{d}{dx}(ye^{\int_a^x P(z)dz}) = Q(x)e^{\int_a^x P(z)dz},$$

where we have used the chain rule of differentiation, the facts that $\frac{d}{dz}e^z = e^z$ and that $\int_a^x P(z)dz$ is an anti-derivative of $P(x)$. Finally, we can once again apply the method of anti-derivatives to find

$$y(x)e^{\int_a^x P(z)dz} = \int_a^x [Q(z)e^{\int_a^z P(t)dt}]dz + C,$$

or

$$y(x) = \int_a^x [Q(z)e^{\int_x^z P(t)dt}]dz + Ce^{-\int_a^x P(z)dz},$$

which is the general solution of the general first-order linear differential equation. It is useful to note that: (i) when Q = 0, the solution reduces to that of the homogeneous equation, y_h; (ii) the general solution is the sum of two parts y_h and $y_p \equiv \int_a^x [Q(z)e^{\int_x^z P(t)dt}dz]$, called the *particular integral.* As we shall see later, these will be the general features for linear second-order differential equations as well.

Examples

(a) Modelling how a person learns: One theory of learning claims that the more a person already knows of the task, the slower he or she learns. In other words, if y represents the percentage of the task that has already been mastered and $\frac{dy}{dt}$ the learning rate per week, then assuming for simplicity that the rate of learning per week is equal to the percentage of task yet to be learned, we get the differential equation

$$\frac{dy}{dt} = 100 - y,$$

or

$$\frac{dy}{dt} + y = 100.$$

By the method of the integrating factor we see that $\frac{d}{dt}(ye^t) = 100e^t$, which we integrate to get $ye^t = 100\int_0^t e^s ds + C'$ or $y = 100 + Ce^{-t}$. If we set the natural initial condition $y = 0$ at $t = 0$, i.e., initially the person started with no knowledge of the task, then the solution is $y = 100(1 - e^{-t})$. This can be represented graphically as in Figure 4.2.

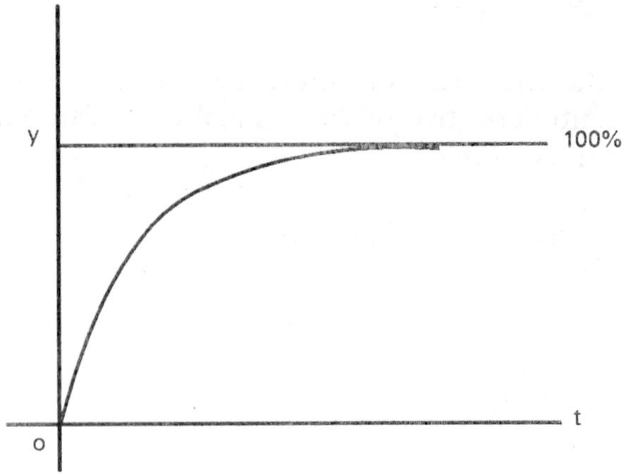

Figure 4.2

(b) Law of heating and cooling: Newton observed from experiments that the temperature of a body changes at a rate proportional to the difference between its temperature and that of its environment. For example, the hotter a cup of tea, the faster it cools. In the long run, the rate of cooling tends to zero and the temperature of the tea approaches that of the environment. If the temperature of the environment is 30° C and that of tea is H (in degree Celsius), then we can express Newton's law of cooling by writing $\frac{dH}{dt} = \alpha(H - 30)$, where α is Newton's cooling constant. If $H > 30$, then the tea must be cooling, i.e., $\frac{dH}{dt}$ must be negative, which implies that α is negative or $\alpha = -k, k > 0$. Thus the equation reads $\frac{dH}{dt} = k(30 - H)$, which is very similar to the equation in the first example and can be solved similarly to get $H = 30 + Ce^{-kt}$. While the constant C can be determined by the initial temperature of the tea (say, 50° C, in which case C turns out to be 20), the cooling constant k can be found by measuring the temperature of the tea at another point in time (**Exercise**). Here we see that though we are studying two very different situations in Examples (a) and (b), the equations and hence the mathematical methods needed to study them are almost identical.

(c) Parabola: Consider the differential equation $y' = 2x$, which we solve by the method of anti-derivatives to get $y = x^2 + C$. For each value of C, we get one parabola so that the general solution $y = x^2 + C$ represents a one-parameter family of parabolas. This is represented in Figure 4.3.

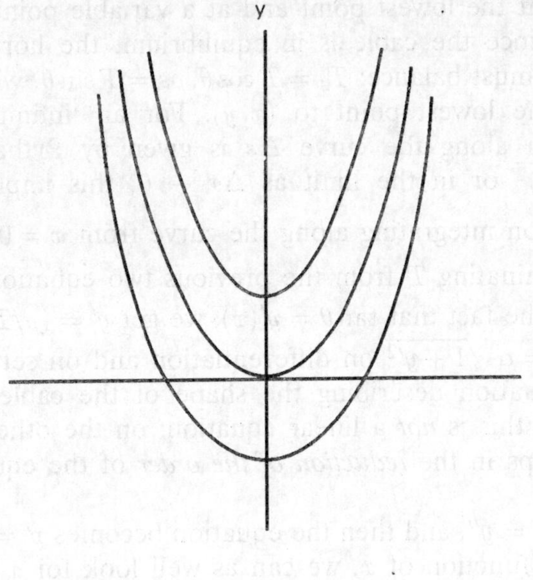

Figure 4.3

(d) The hanging cable: In this example, we choose a problem from Statics which does not involve motion but still results in a differential equation. Consider an inextensible cable of uniform density ρ per unit length, fastened at its two ends and allowed to hang under its own weight (Figure 4.4).

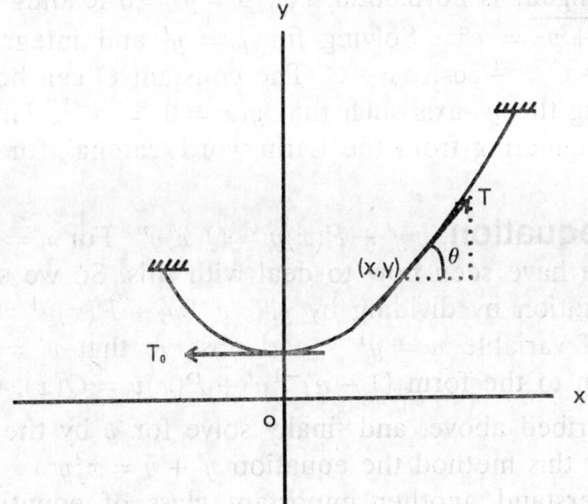

Figure 4.4

We choose the coordinate axes as shown above, and denote by T_0 and T the tension of the cable at the lowest point and at a variable point (x, y) respectively, acting tangentially. Since the cable is in equilibrium, the horizontal and vertical components of these must balance: $T_0 = T\cos\theta, \rho s = T\sin\theta$, where s is the length of the cable from the lowest point to (x, y). For an 'infinitesimal' element of the curve, the length along the curve Δs is given by Pythagorus' theorem as $\Delta s = \sqrt{(\Delta x)^2 + (\Delta y)^2}$ or in the limit as $\Delta x \to 0$; this implies that $(ds/dx) = \sqrt{1 + (dy/dx)^2}$. This on integrating along the curve from $x = 0$ to $x = x$ gives $s = \int_0^x \sqrt{1 + y'(t)^2}dt$. Eliminating T from the previous two equations, using the above expression for s and the fact that $\tan\theta = y'(x)$, we get $y' = (\rho/T_0)\int_0^x \sqrt{1 + y'(t)^2}dt$, which reduces to $y'' = \alpha\sqrt{1 + y'^2}$ on differentiation and on setting $\alpha = \rho/T_0$. This is the differential equation describing the shape of the cable. Now we want to solve it. Observe that this is *not* a linear equation; on the other hand, y does not occur either. This helps in the *reduction of the order* of the equation and finally a solution.

Set $p = y'$ so that $p' = y''$, and then the equation becomes $p' = \alpha\sqrt{1 + p^2}$. Instead of looking for p as a function of x, we can as well look for x as a function of p, and now the equation looks like $\alpha\frac{dx}{dp} = \frac{1}{\sqrt{1+p^2}}$. We can now apply the method of anti-derivatives to get

$$\alpha x = \int_0^p \frac{dt}{\sqrt{1 + t^2}} + C = \log(p + \sqrt{1 + p^2}) + C.$$

The constant of integration C can be obtained by observing that at the lowest point $(x = 0)$ the tangent is horizontal, i.e., $p = y' = 0$ leading to the value $C = 0$ and hence $p + \sqrt{1 + p^2} = e^{\alpha x}$. Solving for $p = y'$ and integrating again we get $y = \frac{1}{2\alpha}(e^{\alpha x} + e^{-\alpha x}) + C = \frac{1}{\alpha}\cosh\alpha x + C$. The constant C can be chosen to be 0 by fixing the scale along the y -axis such that $y(x = 0) = \alpha^{-1}$. The resulting curve is called a *catenary*, originating from the Latin word 'catena', for chain.

(e) Bernoulli's equation: $y' + P(x)y = Q(x)y^n$: For $n = 0, 1$ this reduces to a linear one and we have seen how to deal with this. So we shall assume $n \geq 2$, and rewrite the equation by dividing by $y^n : y^{-n}y' + P(x)y^{1-n} = Q(x)$. Next, we make the change of variable $u = y^{1-n}$ and observe that $u' = (1 - n)y^{-n}y'$. This reduces the equation to the form $(1 - n)^{-1}u' + P(x)u = Q(x)$, which we can solve by the method described above, and finally solve for y by the relation $y = u^{\frac{1}{1-n}}$. (**Exercise**: solve by this method the equation $y' + y = x^2y^2$).

In order to understand another important class of equations of first order, we need to briefly look into the definitions of partial derivatives which we had alluded to in Section 4.2. Let $F : \Delta \subseteq \mathbb{R}^2 \longrightarrow \mathbb{R}$, i.e., a real-valued function of 2 real variables $(x, y) \in \Delta$, a suitable subset of the two-dimensional Euclidean plane. In analogy with the case for one real variable, we call a function F of 2 real variables $(x, y) \in \Delta$ continuous if, given $\varepsilon > 0$, we can find $\delta > 0$ such that $|F(x, y) - F(x_0, y_0)| < \varepsilon$ whenever $\sqrt{(x - x_0)^2 + (y - y_0)^2} < \delta$. We can also talk of derivatives of F with respect to x or y holding y or x respectively constant. If these exist, then we call them the partial derivatives of F with respect to x or y respectively, and write $\frac{\partial F}{\partial x}$ and $\frac{\partial F}{\partial y}$ respectively. More precisely,

$$\frac{\partial F}{\partial x}(x, y) = \lim_{h \to 0} \frac{F(x + h, y) - F(x, y)}{h},$$

$$\frac{\partial F}{\partial y}(x, y) = \lim_{h \to 0} \frac{F(x, y + h) - F(x, y)}{h},$$

if they exist. Of course, $\frac{\partial F}{\partial y}$ and $\frac{\partial F}{\partial x}$ are themselves again functions of x and y and we can consider taking their partial derivatives again and so on. A natural question arises here: Is $\frac{\partial^2 F}{\partial x \partial y} \equiv \frac{\partial}{\partial x}\left(\frac{\partial F}{\partial y}\right)$ equal to $\frac{\partial^2 F}{\partial y \partial x} \equiv \frac{\partial}{\partial y}\left(\frac{\partial F}{\partial x}\right)$? The answer is 'no' in general. However, it can be shown that if the second-order partial derivatives of F exist and are continuous in Δ, then the answer to the above question is 'yes', i.e., $\frac{\partial^2 F}{\partial x \partial y} = \frac{\partial^2 F}{\partial y \partial x}$ (see [1]).

Under small changes in x and y to $x + \Delta x$ and $y + \Delta y$ respectively, $F(x, y)$ changes to $F(x + \Delta x, y + \Delta y)$ and the amount of change

$$\begin{aligned}
\Delta F &\equiv F(x + \Delta x, y + \Delta y) - F(x, y) \\
&= \frac{F(x + \Delta x, y + \Delta y) - F(x, y + \Delta y)}{\Delta x} \Delta x + \frac{F(x, y + \Delta y) - F(x, y)}{\Delta y} \Delta y \\
&= \frac{\partial F}{\partial x} \Delta x + \frac{\partial F}{\partial y} \Delta y + \text{higher orders in } \Delta x \text{ and } \Delta y.
\end{aligned}$$

As in Chapter 3, we write this in the 'differential form'

$$dF = \frac{\partial F}{\partial x} dx + \frac{\partial F}{\partial y} dy.$$

Now let us get back to our topic of differential equations. One comes across a class of equations which can be put in either of the two forms $\frac{dy}{dx} = -M(x, y)/N(x, y)$ (if $N \neq 0$) or $\frac{dx}{dy} = -N(x, y)/M(x, y)$ (if $M \neq 0$). They are actually of the same form except that the roles of x and y and of M and N are interchanged. So without loss of generality, we shall consider only the first form and shall assume furthermore that the functions M and N of the two variables $(x, y) \in \Delta \subseteq \mathbb{R}$ have continuous partial derivatives of the first order in Δ. Often this equation is also written in the so-called *differential form*

$$M(x,y)dx + N(x,y)dy = 0.$$

Finding a solution to this equation would be simple if we can find a function $F : \Delta \to \mathbb{R}$ such that the left hand side of the above equation can be equated to the differential dF of this unknown function. For, in that case, we shall have $dF = 0$, which solves to give $F(x,y) = C$, a constant. Note that in this case, in general we shall only have y as an 'implicit' function of x and not in the form $y = f(x)$. Sometimes, of course, one can algebraically solve the equation $F(x,y) = C$ to find y as an explicit function of x.

The natural question that now arises is: when can we find such an F? If we had such an F, then as we have seen above, $dF = \frac{\partial F}{\partial x}dx + \frac{\partial F}{\partial y}dy \equiv M(x,y)dx + N(x,y)dy$. Since the differentials (increments) dx and dy are independent, it is necessary that

$$M(x,y) = \frac{\partial F}{\partial x} \text{ and } N(x,y) = \frac{\partial F}{\partial y}.$$

Furthermore, since we have assumed at the beginning of this discussion that M and N have continuous first partial derivatives in Δ, it follows by our earlier observation that a necessary condition for the existence of such an F is that M and N satisfy

$$\frac{\partial M}{\partial y}(x,y) = \frac{\partial N}{\partial x}(x,y) \text{ in } \Delta.$$

Immediately the question arises: is this condition also sufficient? Any attempt to answer this question brings us face-to-face with some very interesting and deep 'topological' properties of $\Delta \subseteq \mathbb{R}^2$. First, let us look at the following example: $M(x,y) = -y/(x^2+y^2), N(x,y) = x/(x^2+y^2)$ defined in $\Delta = \mathbb{R}^2 \backslash \{(0,0)\}$, i.e., the Euclidean plane without the origin (or a unit disc with its centre, the origin, scooped out). Clearly M and N are continuous on Δ and also all the partial derivatives of first order exist and are continous. It is also easy to verify that

$$\frac{\partial M}{\partial y}(x,y) = \frac{\partial N}{\partial x}(x,y) = \frac{y^2 - x^2}{(x^2+y^2)^2}.$$

But as we shall soon see, inspite of this there does not exist a function F with $M = \frac{\partial F}{\partial x}$ and $N = \frac{\partial F}{\partial y}$. Assuming to the contrary that such an F exists and changing over to polar coordinates ($x = r\cos\theta, y = r\sin\theta$), we integrate $\frac{\partial F}{\partial \theta}(r\cos\theta, r\sin\theta)$ along a circle of radius 1 to get

$$\int_0^{2\pi} \frac{dF}{d\theta}(\cos\theta, \sin\theta)d\theta = F(1,0) - F(1,0) = 0.$$

On the other hand, the chain rule gives

$$\begin{aligned}
\frac{d}{d\theta}F(\cos\theta, \sin\theta) &= \left.\frac{\partial F}{\partial x}\frac{\partial x}{\partial \theta}\right|_{r=1} + \left.\frac{\partial F}{\partial y}\frac{\partial y}{\partial \theta}\right|_{r=1} \\
&= \frac{\partial F}{\partial x}(-\sin\theta) + \frac{\partial F}{\partial y}\cdot(\cos\theta) \\
&= M(\cos\ \theta, \sin\ \theta)(-\sin\theta) + N(\cos\theta, \sin\theta)\cos\theta \\
&= -(-\sin\theta)\sin\theta + \cos.\theta.\cos\theta = 1
\end{aligned}$$

so that the integral on the left hand side above gives 2π leading to a contradiction and thus establishing our claim that such an F does not exist. This implies that if $\Delta \subseteq \mathbb{R}^2$ has a 'hole' in it, we shall not be able to solve the equation of the form $Mdx + Ndy = 0$ by the above method. Here we shall not be able to answer satisfactorily the question 'for what kind of region $\Delta \subseteq \mathbb{R}^2$ can one solve the above problem', except to state the following theorem. A subset $\Delta \subseteq \mathbb{R}^2$ is said to be *star-shaped* with respect to $x_0 \in \Delta$ if every line segment joining x_0 with every $x \in \Delta$ is in Δ.

THEOREM: *If M and N are functions defined on a star-shaped region $\Delta \subseteq \mathbb{R}^2$ having continuous first partial derivatives in Δ, then there exists a (twice continuously differentiable) function F on Δ such that $dF = Mdx + Ndy$ if and only if $\frac{\partial M}{\partial y} = \frac{\partial N}{\partial x}$.*

For further details, please refer the first chapter of [3].

Examples

(f) $y' = -x/y$. Obviously the equation makes sense only if $y \neq 0$ or away from the x-axis. However, if we convert it to the differential form $xdx + ydy = 0$, it makes sense for all x and y, and we shall consider it only in this form. Thus, $M = x, N = y$ so that $\frac{\partial M}{\partial y} = \frac{\partial N}{\partial x} = 0$ in all of $\mathbb{R}^2$, which is clearly star-shaped. Hence, by the above theorem there must exist an F such that $dF = xdx + ydy$. But in this simple example, we can explicitly write down F, viz., $F(x, y) = \frac{1}{2}(x^2 + y^2)$. Therefore, the solution is $F(x, y) =$ constant, say $c^2/2$ or $x^2 + y^2 = c^2$, which is the equation of a family of concentric circles.

(g) $(2x+e^y)dx + xe^y dy = 0$: Here $M = 2x + e^y$ and $N = xe^y$ so that $\frac{\partial M}{\partial y} = e^y = \frac{\partial N}{\partial x}$ in $\mathbb{R}^2$, and therefore by the above theorem there is an F such that $\frac{\partial F}{\partial x} = M = 2x + e^y$, which we integrate with respect to x to get $F(x,y) = x^2 + xe^y + \varphi(y)$, where $\varphi(y)$ is the constant of integration, treating y as a constant. On the other hand, the second equation $\partial F/\partial y = N$ gives $xe^y + \varphi'(y) = xe^y$ or $\varphi'(y) = 0$, or φ is an absolute constant, say C. This leads to $F(x,y) = x^2 + xe^y + C$ and the solution as $x^2 + xe^y = C_1$, another constant.

4.4 Systems of differential Equations

This is a slightly more advanced topic and we content ourselves with the study of a well-known and beautiful example, viz., that of two interacting populations, the so-called *predator-prey* or *Lotka-Volterra* equation, to get some inkling of what kind of general questions one should ask in this case.

Consider an isolated environment populated only by birds (say myna) and worms; mynas are the predators and worms the prey. Denote by m the population of mynas in thousands and by w, that of worms in millions. If there were no mynas, the worm population would increase exponentially according to the equation $w' = \alpha w$ (α a positive constant). On the other hand, if there were no worms, then the mynas would have no food and would die of starvation so that their population would decrease exponentially according to the equation $m' = -\beta m$ (β a positive constant). If we now have both mynas and worms together, there will be interaction between them. For simplicity, we shall assume that the effect of one population on the other is proportional to the number of 'encounters' between them when a myna eats a worm, and also that the number of encounters is proportional to the product of the two populations since the more there are of either population, the more encounters there will be. So finally we arrive at the system of a pair of equations

$$\frac{dw}{dt} = \alpha w - awm, \quad \frac{dm}{dt} = -\beta m + bwm,$$

where all the constants α, β, a, b are positive. To further simplify the discussion, we look at the case when $\alpha = \beta = a = b = 1$. In such a case, the system simplifies to

$$\frac{dw}{dt} = w(1-m), \quad \frac{dm}{dt} = m(w.-1).$$

This is a non-linear system of equations and cannot usually be solved explicitly. Instead we look at the *phase portrait* of the system, i.e., consider the relation between the two unknowns w and m by eliminating t in the two equations. This we do by using the chain rule and arrive at

$$\frac{dm}{dw} = \frac{dm/dt}{dw/dt} = \frac{m(w-1)}{(1-m)w},$$

which is equivalent to the differential form

$$\left(\frac{1}{m} - 1\right) dm + (1/w - 1)\, dw = 0.$$

This can be integrated by noting that the left hand side above is equal to $d(\log mw - m - w)$ so that $\log(mw) = m + w + c$. Figure 4.5 shows the curve in the m, w phase-plane (for a particular value of c, note that it is a closed curve).

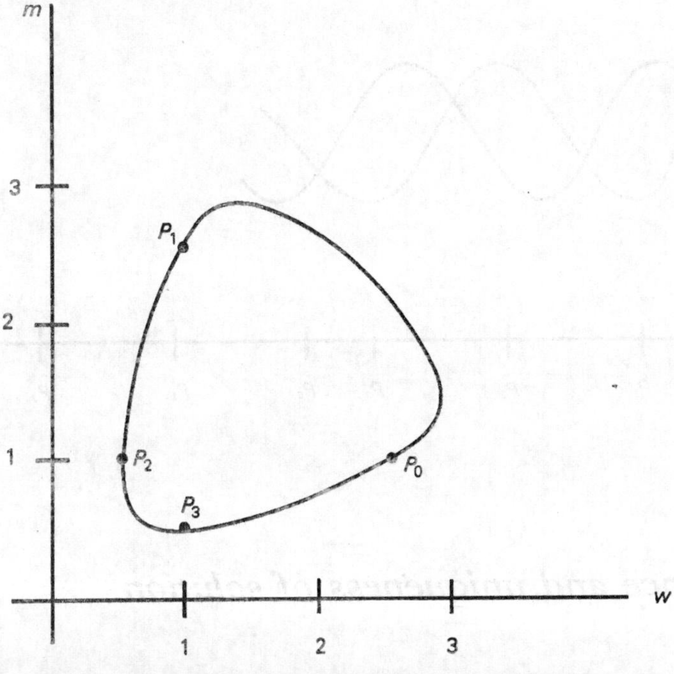

Figure 4.5

The point (1,1) in the phase plane is something special. This is because for $m = w = 1$, both the rates of changes of population $\frac{dm}{dt}$ and $\frac{dw}{dt}$ vanish, giving us the *equilibrium* or *critical point* of this motion. How do we figure out which way to move on the closed trajectory? At P_0 in the Figure 4.5, $m = 1, w > 1$ so that $m' = w - 1 > 0$, while $w' = 0$. Therefore, m is increasing, or the point on the trajectory is moving counter-clockwise.

Finally, what does the shape of the trajectory tell us about how the populations vary with time? If we start at t = 0 at P_0 in Figure 4.5, then move onto P_1 at time

t_1, to P_2 at time t_2, to P_3 at time t_3 and so on, then at some later time, say t_4, we have to be back at P_0 and the whole cycle will repeat. In other words, the closedness of the trajectory in the phase-plane means that both the populations of mynas and of worms will oscillate up and down through the same values (Figure 4.6). Readers interested in learning more in these areas are referred to [4].

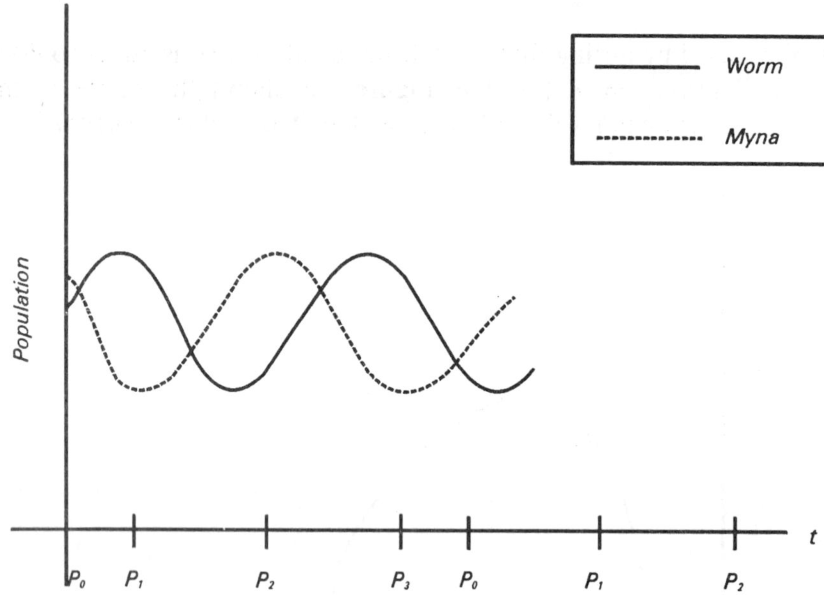

Figure 4.6

4.5 Existence and uniqueness of solution

Once again, this is an advanced topic and we shall only state two theorems without proof, and consider some examples to bring home the ideas. Let $y' = f(y, x)$ be the *normal* form of a first-order differential equation; and we ask: under what condition on f, defined on some region $\Delta \subseteq \mathbb{R}^2$, do solutions to the above equation exist? Intuitively it appears that if f is continuous in both variables in Δ, then we can integrate in small intervals to get the solution. But is the continuity of f enough? That indeed it is, is the content of the next theorem.

THEOREM (PEANO'S EXISTENCE THEOREM): *Let $y' = f(y, x)$ with f continuous in $\Delta \subseteq \mathbb{R}^2$, and let $(x_0, y_0) \in \Delta$. Then there exists an interval I about x_0 and at least one function $y = y(x)$ defined on I such that*

(i) $y(x_0) = y_0$ *and*
(ii) $y'(x) = f(x, y(x))$ *for all* $x \in I$.

Note that the theorem contains no assertion about the uniqueness of the solution $y = y(x)$. In fact, without further restrictions on the function f, the above equation need not have an unique solution as can be seen from the following example.

Consider the initial value problem

$$y' = 3y^{2/3}, \quad y(-2) = -1.$$

Here $f(y, x) = 3y^{2/3}$ is a continuous function on all of $\mathbb{R}^2$ and hence by the above theorem, there exists at least one solution. Indeed, rewriting the equation as $y^{-2/3}dy = 3dx$ and integrating, we get $y^{1/3} = x + c$ or $y = (x + c)^3$. The condition $y(-2) = -1$ leads to $c = 1$ or the solution $y = (x+1)^3$. But note that $y = 0$ is also a solution of the equation though it does not satisfy the intitial value. Many solutions to the initial value problem can be found by 'patching together' $y = (x + c)^3$ and $y = 0$. For example,

$$y = \begin{cases} (x+1)^3 & \text{if } x \le -1 \\ 0 & \text{if } x > -1, \end{cases}$$

and

$$y = \begin{cases} (x+1)^3 & \text{if } x \le -1 \\ 0 & \text{if } -1 < x < a \\ (x-a)^3 & \text{if } x \ge a, \end{cases}$$

(where $a > -1$ is an arbitrary constant), give a one-parameter family of solutions to the above differential equation. The kind of further sufficient condition that one needs on f to ensure uniqueness of the solution is given in the following theorem.

THEOREM (PICARD'S UNIQUENESS THEOREM): *Let* $y' = f(y, x)$ *with* f *satisfying a Lipschitz condition, i.e.,* $|f(y_1, x) - f(y_2, x)| \le k|y_1 - y_2|$ *for some constant* k *and all* x *in some closed interval* I. *Then there exists a unique solution to the initial value problem* $y' = f(y, x)$; $y(x_0) = y_0$ *with* $x_0 \in I$.

In the above example, the estimate $|y_1^{2/3} - y_2^{2/3}| \le k|y_1 - y_2|$ breaks down near $y = 0$ and thus Picard's theorem cannot be applied.

As a simple application of this theorem, let us consider the linear first-order equation $y' + P(x)y = Q(x)$ (with P and Q continuous in the interval I) to which we already have a solution. We now ask: is the solution the only one? If we cast this equation in the form as in the second theorem above, then we find that $f(y, x) = Q(x) - P(x)y$. Thus,

$$\begin{aligned}|f(y_1,x)-f(y_2,x)| &= |P(x)(y_2-y_1)|\\ &\leq \sup_{x\in I}|P(x)|\ |y_1-y_2| = k|y_1-y_2|,\end{aligned}$$

where we have set $k=\sup\limits_{x\in I}|P(x)|$ (which is finite since P is continuous in I, see Chapter 3). Therefore the hypothesis of Picard's theorem is satisfied and we are now sure that the solution we had found earlier is indeed the only one.

4.6 Second-order differential equation

In this section, we shall occasionally refer to some of the material on linear mappings covered in Chapter 5. At this stage, the reader can either postpone the reading of this section till he/she finishes Chapter 5, or continue with the present section assuming the ideas from Chapter 5.

We have already come across a second-order (linear) differential equation while discussing in Section 4.1 the motion of a body under the influence of gravity. There the solution, the distance s, is given as $s=-\frac{1}{2}gt^2+v_0t+s_0$, and we see that it contains two constants of integration, viz., the initial position $s(0)=s_0$ and initial velocity $s'(0)=v_0$. In other words, the general solution is a two-parameter family. As we shall see later, this is a general feature of all second-order linear differential equations and not only of the ones with constant coefficients as was the above equation. Here we shall restrict our attention only to the linear case; in fact, mostly to the ones with constant coefficients.

Consider a general linear equation of second order: $a_0(x)y''+a_1(x)y'+a_2(x)y=g(x)$, where the coefficient functions a_0,a_1,a_2 and g are assumed to be continuous in an interval $I\subseteq\mathbb{R}$, and a_0 is assumed not vanishing anywhere in I. Thus by dividing by a_0 we can consider instead: $y''+p_1(x)y'+p_2(x)y=q(x)$, where the functions p_1,p_2 and q are continuous in I. We shall now rewrite this as a system of a pair of first-order differential equations and then re-interpret that as a first-order linear differential equation where the unknown function takes values in $\mathbb{R}^2$, i.e., it is a two-dimensional vector-valued function of a real variable (recall the definition of a general function from Chapter 1 and the reader is encouraged to digress here for a while to read the first few pages of Chapter 5 for the concept of a two-dimensional vector space). Thus the values of such a function have two components or conversely, such a function is completely described by specifying its two component functions.

Let $z: I\subseteq\mathbb{R}\to\mathbb{R}^2$ be given by its two components $z_1(x)=y(x)$ and $z_2(x)=y'(x)$ for all $x\in I$. Then the above second-order equation can be rewritten as

$$z_1'(x)=y'(x)=z_2(x)$$

$$z_2'(x) = y''(x) = -p_1(x)y' - p_2(x)y + q(x) = -p_1(x)z_2(x) - p_2(x)z_1(x) + q(x)$$

which can be combined in matrix notation as (the reader is again referred to Chapter 5 for this)

$$\begin{pmatrix} z_1'(x) \\ z_2'(x) \end{pmatrix} = \begin{pmatrix} 0 & 1 \\ -p_2(x) & -p_1(x) \end{pmatrix} \begin{pmatrix} z_1(x) \\ z_2(x) \end{pmatrix} + \begin{pmatrix} 0 \\ q(x) \end{pmatrix}.$$

If we now set

$$A(x) = \begin{pmatrix} 0 & 1 \\ -p_2(x) & -p_1(x) \end{pmatrix},$$

a 2×2 matrix-valued function on I and $\boldsymbol{q}(x) = \binom{0}{q(x)}$, a 2-dimensional vector represented in the form of a column matrix, then the above can be recast in the compact form

$$\frac{d\boldsymbol{z}}{dx}(x) = A(x)\boldsymbol{z}(x) + \boldsymbol{q}(x).$$

In other words, a typical (normal) second-order linear equation has been reinterpreted as a linear first-order equation at the cost of making the unknown function into one taking values in a 2-dimensional vector space. Here we must remark that though we have stated Picard's theorem in the context of the unknown function (the solution) taking only real values, the theorem remains valid even when the solution y takes values in a large class of vector spaces, in particular, for 2-dimensional vector spaces. Taking this for granted, it is clear that, as in the last part of Section 4.5,

$$\|(A(x)\boldsymbol{z}^{(1)}(x) + \boldsymbol{q}(x)) - (A(x)\boldsymbol{z}^{(2)}(x) + \boldsymbol{q}(x))\| = \|A(x)(\boldsymbol{z}^{(1)}(x) - \boldsymbol{z}^{(2)}(x))\|$$

$$= \|A(x)\gamma\| \qquad \text{(we have set } \gamma = \boldsymbol{z}^{(1)} - \boldsymbol{z}^{(2)})$$

$$= \left\| \begin{pmatrix} \gamma_2 \\ -p_1\gamma_1 - p_2\gamma_2 \end{pmatrix} \right\| = \sqrt{\gamma_2^2 + (p_1\gamma_1 + p_2\gamma_2)^2}.$$

Here we have used the specific form of the matrix A and a measure of the size of a 2-dimensional vector γ as the norm $\|\gamma\| = (\gamma_1^2 + \gamma_2^2)^{\frac{1}{2}}$ (see Section 5.8). Thus the right hand side above is less than or equal to $K(|\gamma_1| + |\gamma_2|) = K\|\boldsymbol{z}^{(1)} - \boldsymbol{z}^{(2)}\|$, where $K = \max(1, \sup_{x \in I}|p_1(x)|, \sup_{x \in I}|p_2(x)|)$.

Thus we conclude that the vector differential equation with initial value $z(x_0) = z^{(0)}$ for some $x_0 \in I$ has a unique solution. Recalling the definition of the vector function z, this implies that the (normal) second-order linear differential equation that we started with, along with the initial values $y(x_0) \equiv z_1(x_0) = z_1^{(0)}$ and $y'(x_0) \equiv z_2(x_0) = z_2^{(0)}$, has a unique solution.

In the rest of this section we shall further specialize to the case when the coefficient functions a_0, a_1 and a_2 are constants with $a_0 \neq 0$. Since constant functions are continuous, we know from the last paragraph that the equation

$$a_0 y'' + a_1 y' + a_2 y = q(x)$$

has a unique solution. Our next task is to find one solution which, by virtue of the uniqueness of the solution, will be the *only* solution. This aim will be achieved in two steps: (i) we shall first find the general solution y_h of the associated homogeneous equation $a_0 y'' + a_1 y' + a_2 y = 0$ and then (ii) find any *one* solution y_p (called the *particular integral*) for the inhomogeneous one. After this we note that because of the linearity of the equation, $y = y_h + y_p$ is a solution of the original equation. In fact, we have the following interesting but elementary theorem.

THEOREM: *There exist functions y_1, y_2 such that the set of solutions of the homogeneous equation $a_0 y'' + a_1 y' + a_2 y = 0$ in I is given by $\{\alpha y_1 + \beta y_2 | \alpha, \beta \in \mathbb{R}\}$. Furthermore, for any $\alpha, \beta \in \mathbb{R}$ the function $\alpha y_1 + \beta y_2$ is not identically zero unless $\alpha = \beta = 0$ (i.e., y_1 and y_2 are linearly independent;* see Section 5.4*) and this fact is expressed as: the set of solutions of the equation form a two-dimensional vector space.*

PROOF: Let x_0 be a fixed point in I; then we know that the above equation admits a pair of unique solutions y_1 and y_2 satisfying the corresponding sets of initial values $y_1(x_0) = 1$, $y_1'(x_0) = 0$ and $y_2(x_0) = 0$, $y_2'(x_0) = 1$ respectively. First of all, we want to verify that these two solutions y_1 and y_2 are linearly independent (see Chapter 5) and then confirm that every solution y of the equation is a linear combination of y_1 and y_2. For the first part, assume that $\alpha_1 y_1(x) + \alpha_2 y_2(x) = 0$ for all $x \in I$. By differentiating this, we get $\alpha_1 y_1'(x) + \alpha_2 y_2'(x) = 0$ also for all $x \in I$. Substituting the initial values for y_1, y_1', y_2 and y_2' we get $\alpha_1 = \alpha_2 = 0$, i.e., that y_1 and y_2 are linearly independent. For the second part, let y be the solution with initial values $y(x_0) = \alpha$ and $y'(x_0) = \beta$. But it is clear that the function $\alpha y_1 + \beta y_2$ is also a solution of the equation having the same initial values. The conclusion now follows from the uniqueness of the initial value problem as discussed earlier. ■

Thus the step (i) of finding a solution of the homogeneous equation has been reduced to finding two linearly independent solutions and then taking linear combinations of the two. Let us start with a simple example $y'' - 4y = 0$, and try the solution $y(x) = e^{mx}$. On substitution this yields $(m^2 - 4)e^{mx} = 0$ for all $x \in \mathbb{R}$. It should be apparent to the reader that the exponential function of a real (or even a complex) variable vanishes nowhere and therefore m must satisfy $m^2 - 4 = 0$. This simple quadratic has two distinct roots, viz., $m = \pm 2$ leading to the conclusion that $y_1(x) = e^{2x}$ and $y_2(x) = e^{-2x}$ are two solutions. They are linearly independent since, though $y_1(0) = y_2(0) = 1$, $y_1'(0) = 2$ while $y_2'(0) = -2$ (check it up !). Thus the general solution to the equation $y'' - 4y = 0$ is $y(x) = c_1e^{2x} + c_2e^{-2x}$, where c_1 and c_2 are two constants to be determined by appropriate initial values.

In the general case, one again uses the trial solution $y = e^{mx}$ to get the quadratic equation (called the *characteristic equation*) $a_0m^2 + a_1m + a_2 = 0$ ($a_0 \neq 0$) which can be solved to get two roots (they may be complex, e.g., in the case of the equation $y'' + 4y = 0$ leading to $m^2 + 4 = 0$ which has two solutions $m = \pm 2i$; here we are assuming some elementary knowledge of complex numbers). There arise three different cases.

Case I (the two roots m_1 and m_2 are real and unequal): This is the case we have dealt with in the beginning. Here e^{m_1x} and e^{m_2x} are two linearly independent solutions (**Exercise !**) so that the general solution is $y(x) = c_1e^{m_1x} + c_2e^{m_2x}$.

Case II (m_1 and m_2 are real and equal to $\bar{m}$, say): Now we have the problem that the function $e^{\bar{m}x}$ gives just *one* solution. How do we find the second solution, linearly independent of $e^{\bar{m}x}$? It is easy to verify that

$$\begin{aligned}
& a_0(xe^{\bar{m}x})'' + a_1(xe^{\bar{m}x}) + a_2xe^{\bar{m}x} \\
&= a_0\{2\bar{m}e^{\bar{m}x} + \bar{m}^2xe^{\bar{m}x}\} + a_1\{e^{\bar{m}x} + \bar{m}xe^{\bar{m}x}\} + a_2xe^{\bar{m}x} \\
&= e^{\bar{m}x}(a_1 + 2a_0\bar{m}) + xe^{\bar{m}x}(a_0\bar{m}^2 + a_1\bar{m} + a_2) \\
&= 0,
\end{aligned}$$

since $a_0(m - \bar{m})^2 = a_0m^2 + a_1m + a_2$ for all m implies that $-2a_0\bar{m} = a_1$. Thus $y_2(x) = xe^{\bar{m}x}$ is another solution and is also linearly independent of $y_1(x) = e^{\bar{m}x}$. Therefore, the general solution in this case is $y(x) = c_1y_1(x) + c_2y_2(x) = (c_1 + c_2x)e^{\bar{m}x}$.

Case III (m_1 and m_2 are non-real, complex conjugate to each other): Let $m_1 = \alpha + i\beta$, and $m_2 = \alpha - i\beta$, α, β real and $\beta \neq 0$. Now, one can directly verify (**Exercise!**) that $y_1(x) = e^{\alpha x}\cos\beta x$ and $y_2(x) = e^{\alpha x}\sin\beta x$ are two solutions of the differential equation in this case. That they are linearly independent essentially follows from the fact that $\cos\beta x$ and $\sin\beta x$ are linearly independent. This can be seen as follows. Let $c_1\cos\beta x + c_2\sin\beta x = 0$ for all x. Then by differentiating, we get $\beta(c_2\cos\beta x - c_1\sin\beta x) = 0$. Now setting $x = 0$ and remembering that $\beta \neq 0$, we have $c_1 = c_2 = 0$ proving what we wanted to prove.

Remarks: (a) One can deal with case III in another way, without using complex numbers. We give a sketch and encourage the reader to complete the process. Look at the equation $y'' + 2by' + \omega^2 y = 0$. First substitute $y = e^{-bx}u$ to get $u'' + (\omega^2 - b^2)u = 0$, or $u'' + \omega_0^2 u = 0$, where $\omega_0^2 = \omega^2 - b^2 > 0$. Next set $v = (du/dx)$ so that $u'' = (dv/dx) = v(dv/du)$. This transforms the equation into $v(dv/du) + \omega_0^2 u = 0$ or $(d/du)(v^2 + \omega_0^2 u^2) = 0$ leading to the conclusion $v^2 + \omega_0^2 u^2$ = a constant, say $\omega_0^2 a^2$, or $v = \pm\omega_0\sqrt{a^2 - u^2}$ or $(du/\sqrt{a^2 - u^2}) = \pm\omega_0 dx$. This we can integrate to get $\sin^{-1}(u/a) = \pm\omega_0 x + \theta$.

(b) There is yet another way of considering the equation in (a), one in which we make contact with the concepts of linear mappings in $\mathbb{R}^2$ as introduced in Chapter 5.

Set $p = \omega_0^{-1}(y' + by)$ and rewrite the equation as a pair of first-order equations in terms of x and p (as explained in the beginning of this section):

$$y' = \omega_0 p - by, \quad p' = -bp - \omega_0 y.$$

Now if we write $z(x) = \begin{pmatrix} y(x) \\ p(x) \end{pmatrix}$, then z satisfies the first-order equation

$$z'(x) = \begin{pmatrix} -b & \omega_0 \\ -\omega_0 & -b \end{pmatrix} z(x) = (-bI + \omega_0\sigma)z(x),$$

where I is the 2×2 identity matrix and $\sigma = \begin{pmatrix} 0 & 1 \\ -1 & 0 \end{pmatrix}$. As in the one-dimensional case, this can be solved:

$$z(x) = \exp(x(-bI + \omega_0\sigma))z_0 = e^{-bx}\exp(x\omega_0\sigma)z_0.$$

If we now consider $\exp(x\omega_0\sigma)$ as given by its (infinite) series expansion and note that σ^n is $(-1)^{n/2}I$ for even n and is $(-1)^{(n-1)/2}\sigma$ for odd n, we get

$$\begin{aligned} \exp(x\omega_0\sigma) &= I \cdot \cos x\omega_0 + \sigma \cdot \sin x\omega_0 \\ &= \begin{pmatrix} \cos x\omega_0 & \sin x\omega_0 \\ -\sin x\omega_0 & \cos x\omega_0 \end{pmatrix}. \end{aligned}$$

Next, putting all these together (verify!) one gets the expected result

$$\begin{aligned} y(x) &= \{y(0)\cos x\omega_0 + p(0)\sin x\omega_0\}e^{-bx} \\ &= y(0)e^{-bx}\{\cos x\omega_0 + \tfrac{b}{\omega_0}\sin x\omega_0\} + \tfrac{y'(0)}{\omega_0}e^{-bx}\sin x\omega_0. \end{aligned}$$

Examples

(g) Spring and Hooke's law: Most of us have experienced the following feature of springs. With reference to Figure 4.7, we expect that the farther the body of mass m is from the equilibrium position, i.e., the more the spring is stretched or compressed, the larger is the force and also that the force of restitution is directed against the act of stretching or compression of the spring. In fact, it is experimentally established that if the displacement is not too large to deform the spring permanently, then the force is proportional to the displacement, and is directed opposite to it, i.e.,

$$F = -kx,$$

where k is the spring constant ($k > 0$) which depends on the material of the spring. This is *Hooke's law*. On the other hand, Newton's second law of motion says that this force would give rise to an acceleration (d^2x/dt^2) such that

$$m\frac{d^2x}{dt^2} = -kx \quad \text{or} \quad \frac{d^2x}{dt^2} + \left(\frac{k}{m}\right)x = 0.$$

The associated characteristic equation is $m^2 + \omega^2 = 0$, where $\omega = \sqrt{k/m}$, called the characteristic frequency. This equation has two complex solutons $m = \pm i\omega$.

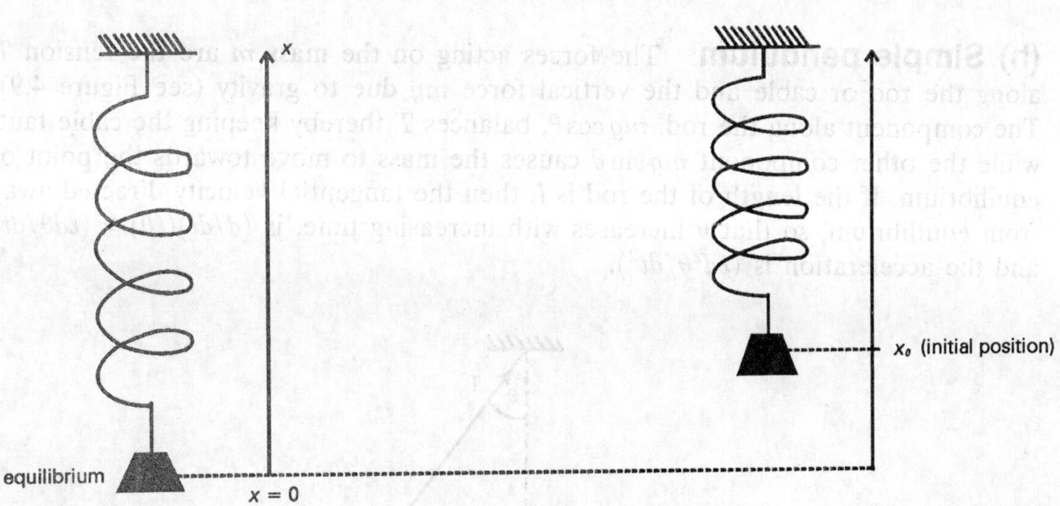

Figure 4.7

By what we have discussed above in case III, the general solution is $x(t) = c_1 \cos\omega t + c_2 \sin\omega t$. If now the physical situation is such that the spring is compressed (as in Figure 4.7) to the position $x(0) = x_0$ and then released so that the initial velocity $x'(0) = 0$, then we can determine the constants c_1 and c_2; $c_1 = x_0$ and $\omega c_2 = 0$ which lead to the solution of the initial value problem as $x(t) = x_0 \cos \omega t$.

In general, c_1 and c_2 can instead be expressed in the polar form $c_1 = a \sin \theta$ and $c_2 = a \cos \theta$, and this leads to an equivalent but convenient expression for the solution $x(t) = a(\cos \omega t \sin \theta + \sin \omega t \cos \theta) = a \sin(\omega t + \theta)$. The motion looks as in Figure 4.8.

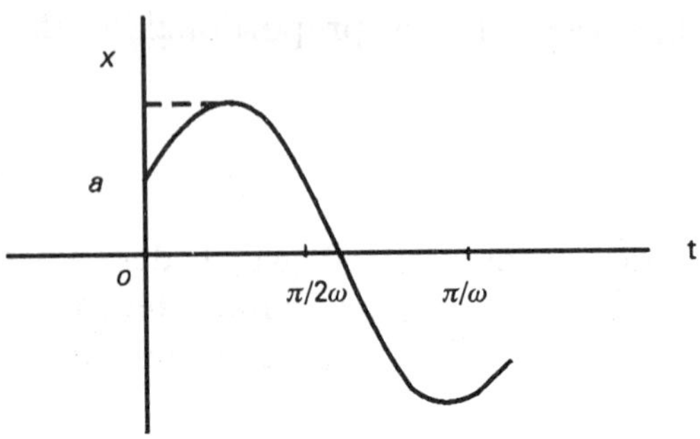

Figure 4.8

The maximum value a of $x(t)$ is called the *amplitude* and θ $(-\pi \leq \theta \leq \pi)$ is called the *phase* of the motion. The motion is clearly oscillatory or periodic in time.

(h) Simple pendulum: The forces acting on the mass m are the tension T along the rod or cable and the vertical force mg due to gravity (see Figure 4.9). The component along the rod, $mg \cos \theta$, balances T thereby keeping the cable taut, while the other component $mg \sin \theta$ causes the mass to move towards the point of equilibrium. If the length of the rod is l, then the tangential velocity directed away from equilibrium, so that θ increases with increasing time, is $(d/dt)(\ell\theta) = (\ell d\theta/dt)$ and the acceleration is $(\ell d^2\theta/dt^2)$.

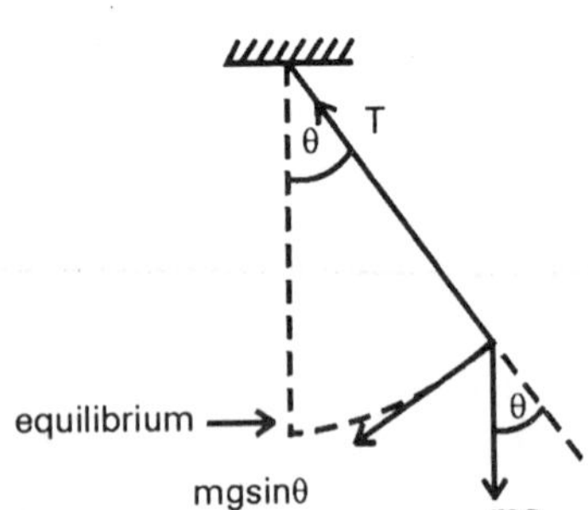

Figure 4.9

Thus the equation of motion of this system (called a *simple pendulum*) is

$$m\ell\frac{d^2\theta}{dt^2} = -mg\sin\theta$$

or $$\frac{d^2\theta}{dt^2} + \left(\frac{g}{\ell}\right)\sin\theta = 0.$$

For small displacements, i.e., for small θ, we can approximate $\sin\theta$ by θ and the equation takes the form

$$\frac{d^2\theta}{dt^2} + \left(\frac{g}{\ell}\right)\theta = 0,$$

which looks exactly like the one we had derived for a spring in Example (g) and admits a solution exactly as was shown there. This again goes to show that many situations in applications lead to identical equations and hence no new mathematics.

(i) L-C-R electrical circuit: A resistor R, an inductance L and a capacitor C are connected in series with a source of electromotive force (e.m.f.) E (see Figure 4.10). If I is the current flowing through the circuit in equilibrium, the potential drop across R is RI and across the inductance is $L(dI/dt)$. A capacitor consists of two conducting plates separated by a thin layer of insulator and when an e.m.f. e is applied across it, it accumulates charge Q such that $e = Q/C$, where C is called the *capacitance* and depends on the characteristics of the insulator. Combining all these, we get for the equilibrium of the circuit:

$$L\frac{dI}{dt} + RI + \frac{Q}{C} = E.$$

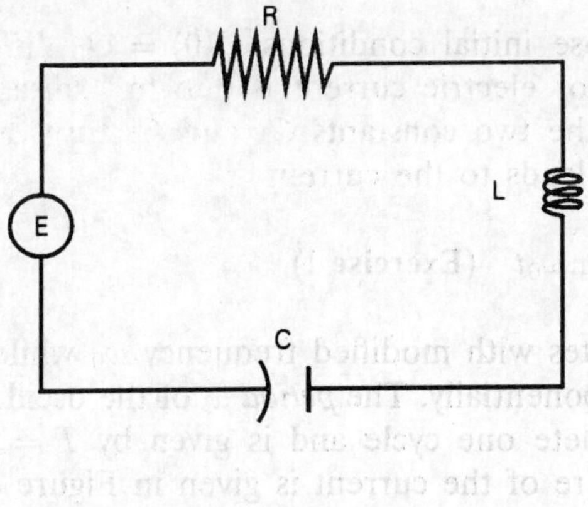

Figure 4.10

On the other hand, electric current I is just the rate of change of the accumulated charge, i.e., $I = (dQ/dt)$ which allows us to rewrite the above as

$$L\frac{d^2Q}{dt^2} + R\frac{dQ}{dt} + \frac{Q}{C} = E.$$

Now the equation looks familiar. In the absence of any external e.m.f (i.e., $E = 0$), we have the homogeneous equation $L(d^2Q/dt^2) + R(dQ/dt) + (1/C)Q = 0$. The associated characteristic equation is $Lm^2 + Rm + C^{-1} = 0$. As before, three cases arise. If $R = 0$, then the general solution for charge Q as a function of time is $Q(t) = C_1 \cos \omega t + C_2 \sin \omega t$, where $\omega = (LC)^{-1/2}$, giving rise to the kind of oscillatory motion of charges as seen earlier.

Next we consider the case when $R \neq 0$ and set $\omega^2 = (LC)^{-1}$ as before and $2b = RL^{-1}$, leading to the equation $Q''+2bQ'+\omega^2 Q = 0$. We have three different situations according as $b > \omega$, $b = \omega$, $b < \omega$. When $b > \omega$, both the roots m_1 and m_2 of the associated characteristic equation are real and unequal ($m_1 = -b + \sqrt{b^2 - \omega^2}, m_2 = -b - \sqrt{b^2 - \omega^2}$), and the solution is $Q(t) = e^{-bt}(C_1 e^{t\sqrt{b^2-\omega^2}} + c_2 e^{-t\sqrt{b^2-\omega^2}})$. Since $\sqrt{b^2 - \omega^2} < b$, it is clear that in this case, $Q(t) \to 0$ as $t \to \infty$. Thus if $R^2C > 4L$ (i.e., if $b^2 > \omega^2$), we have the *overdamped* or *transient* charges and currents.

If on the other hand $b < \omega$ (i.e., $R^2C < 4L$), the two roots of the characteristic equation are complex ($m_1 = -b + i\sqrt{\omega^2 - b^2}, m_2 = -b - i\sqrt{\omega^2 - b^2}$) and the solution is given by

$$\begin{aligned} Q(t) &= e^{-bt}(C_1 \cos t\sqrt{\omega^2 - b^2}) + C_2 \sin(t\sqrt{\omega^2 - b^2})) \\ &= e^{-\frac{tR}{2L}}[C_1 \cos \omega_0 t + C_2 \sin \omega_0 t], \end{aligned}$$

where $\omega_0 = \sqrt{\omega^2 - b^2} = [1/LC - (R/2L)^2]^{1/2}$, and the current I in the circuit has the expression

$$I(t) = \frac{dQ}{dt} = e^{-\frac{tR}{2L}}\left[(\omega_0 C_2 - \frac{R}{2L}C_1)\cos \omega_0 t - (\omega_0 C_1 + \frac{R}{2L}C_2)\sin \omega_0 t\right]$$

In this case if we impose initial conditions $Q(0) = Q_0, I(0) = 0$, which describe a circuit in which flow of electric current is due to a discharging capacitor with initial charge Q_0, then the two constants C_1 and C_2 turn out to be $C_1 = Q_0$ and $C_2 = RQ_0/(2L\omega_0)$. This leads to the current

$$I(t) = \frac{-Q_0}{LC\omega_0}e^{-\frac{tR}{2L}} \sin \omega_0 t \quad \textbf{(Exercise !)}.$$

The current now oscillates with modified frequency ω_0 while at the same time its magnitude decreases exponentially. The *period* T of the oscillation is the time taken by the current to complete one cycle and is given by $T = 2\pi/\omega_0 = 2\pi[(LC)^{-1} - (R^2/4L^2)]^{-1/2}$. The nature of the current is given in Figure 4.11.

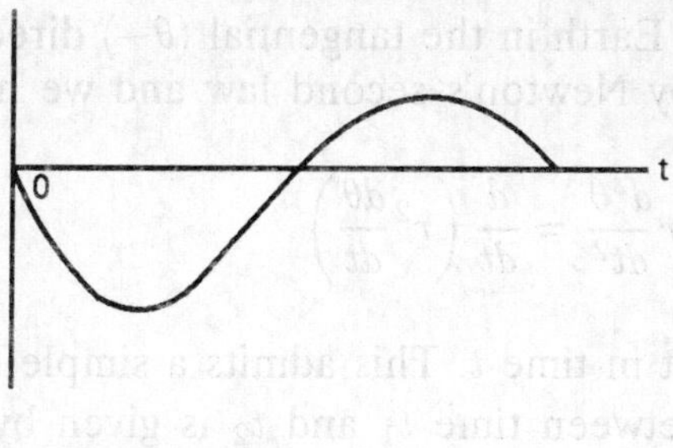

Figure 4.11

We do not discuss the case when $b=\omega$ and leave it as an exercise. Also the situation when $E \neq 0$ will be considered later. The equations of the type: $y''+2by'+\omega^2 y=g$ (the so-called *forced and damped harmonic oscillator*) appear very often in many different situations, but the basic mathematical discussion (when $g=0$) remains as above.

(j) Planetary motion and Kepler's laws: We picture the Sun of mass M attracting earth of mass m with the force of attraction given by $F=-(GMm)/r^2$, where r is the distance between the Sun and Earth, and G is Newton's gravitational constant (see Figure 4.12). Strictly speaking, this is a two-body problem, but for simplicity (which is justified, as the mass of the sun is much larger as compared to that of Earth), we shall imagine the Sun to be fixed while Earth moves around. Since the force is a function of only the distance r between the Sun and Earth, it is clear that the motion (though in 3-dimensional physical space) is actually taking place in the plane swept out by r and thus it is convenient to rewrite the problem in the 2-dimensional polar (r,θ)-coordinate system $x=r\cos\theta, y=r\sin\theta$. It can be shown (some of the readers may have seen this in their physics class) that the accelerations in the radial and angular directions are given respectively by

$$a_r=\frac{d^2r}{dt^2}-r\left(\frac{d\theta}{dt}\right)^2 \quad \text{and} \quad a_\theta=2\left(\frac{dr}{dt}\right)\left(\frac{d\theta}{dt}\right)+r\frac{d^2\theta}{dt^2}.$$

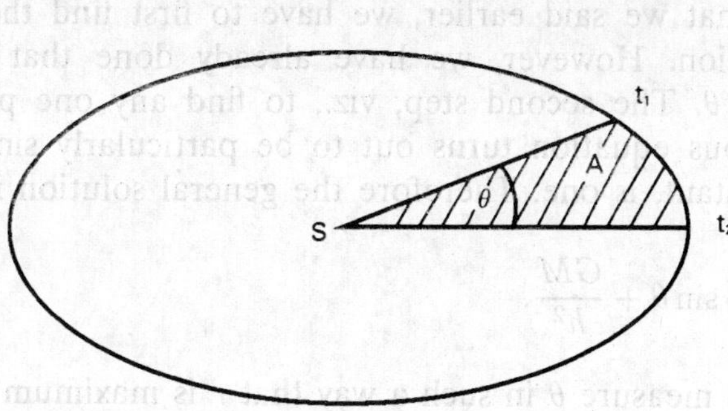

Figure 4.12

Since there is no force on Earth in the tangential ($\theta-$) direction, the corresponding acceleration a_θ vanishes by Newton's second law and we have

$$0 = 2\left(\frac{dr}{dt}\right)\left(\frac{d\theta}{dt}\right) + r\frac{d^2\theta}{dt^2} = \frac{d}{dt}\left(r^2\frac{d\theta}{dt}\right)$$

or, $r^2(d\theta/dt) = h$, constant in time t. This admits a simple interpretation. The area A, swept out by Earth, between time t_1 and t_2 is given by $A|_{t_1}^{t_2} = (1/2)\int_{t_1}^{t_2} r^2 d\theta = (1/2)\int_{t_1}^{t_2} h dt = (1/2)h(t_2 - t_1)$, i.e., the radius vector from the Sun to Earth sweeps out equal areas in equal time intervals. This is *Kepler's second law* of planetary motion, established by Kepler from the observations of the astronomer Tycho Brahe.

What about the radial part of the motion? Applying Newton's second law for the radial part we have

$$ma_r \equiv m\left[\frac{d^2r}{dt^2} - r\left(\frac{d\theta}{dt}\right)^2\right] = -\frac{GMm}{r^2}.$$

A change of variable $u = r^{-1}$ helps in solving this equation. Since $\frac{du}{dt} = -r^{-2}\frac{dr}{dt}$, $d^2r/dt^2 = \frac{d}{dt}\left(-r^2\frac{du}{dt}\right) = -\frac{d}{dt}\left(r^2\frac{d\theta}{dt}\cdot\frac{du}{d\theta}\right) = -\frac{d}{dt}\left(h\frac{du}{d\theta}\right) = -h\frac{d}{d\theta}\left(\frac{du}{d\theta}\right)\cdot\frac{d\theta}{dt} = -h^2r^{-2}\frac{d^2u}{d\theta^2} = -h^2u^2\frac{d^2u}{d\theta^2}$, as $r^2\frac{d\theta}{dt} = h$, a constant. Similarly, $r\left(\frac{d\theta}{dt}\right)^2 = \left(r^2\frac{d\theta}{dt}\right)^2 r^{-3} = u^3h^2$ and the equation is rewritten as

$$-mh^2\left[u^2\frac{d^2u}{d\theta^2} + u^3\right] = -GMmu^2$$

or,

$$\frac{d^2u}{d\theta^2} + u = GM/h^2.$$

But according to what we said earlier, we have to first find the solution of the homogeneous equation. However, we have already done that in Example (g): $u_h = c_1\cos\theta + c_2\sin\theta$. The second step, viz., to find any one particular solution of the inhomogeneous equation turns out to be particularly simple in this case: $u_p = GM/h^2$, a constant, is one. Therefore the general solution is

$$u = c_1\cos\theta + c_2\sin\theta + \frac{GM}{h^2}.$$

If we now choose to measure θ in such a way that r is maximum when $\theta = 0$, then it follows (see discussions on maxima–minima in Section 4.7) that $(du/d\theta) = 0$ and

$d^2u/d\theta^2 < 0$ when $\theta = 0$. This implies that $c_2 = 0$ and $c \equiv c_1 > 0$, and the orbit of Earth traces out a curve in the Sun–Earth plane given by

$$r = \frac{h^2/GM}{1 + (ch^2/GM)\cos\theta},$$

which is the polar equation of an ellipse with eccentricity (ch^2/GM). This is *Kepler's first law*.

In the last example, we have actually solved an inhomogeneous second-order linear differential equation by inspection and it worked because the inhomogeneity was particularly simple, viz., a constant. In the last part of this section we shall see briefly how one deals with this problem in general. Since here we are mostly concerned with equations with constant coefficients, the method of *undetermined coefficients* that we are going to describe is the most efficient. However, there is the more widely applicable method of *Green's functions* which we shall not describe here.

Consider the differential equation $L_1 y \equiv a_0 y'' + a_1 y' + a_2 y = g$, where L stands for the *differential* operator $a_0(d^2/dx^2) + a_1(d/dx) + a_2$. Also assume that the inhomogeneity g itself is a solution of another linear (constant coefficient) differential operator L_2, in which case we get $L_2 L_1 y = 0$ or y as the solution of a higher-order constant-coefficient differential equation. This we can often solve by factorizing the associated characteristic equation, the solution being given by a linear combination of various characteristic solutions; the coefficients are finally determined by actually substituting it into the given equation. This is best demonstrated by examples, say Example (i) for an L-C-R circuit with e.m.f. $E = E_0 \sin\bar{\omega}t$, $\bar{\omega}$ being the frequency of the impressed alternating current source.

The equation we are looking at is

$$\frac{d^2Q}{dt^2} + 2b\frac{dQ}{dt} + \omega^2 Q = e_0 \sin\bar{\omega}t, \quad \text{where } e_0 = (E_0/L).$$

We know that $g(t) \equiv e_0 \sin\bar{\omega}t$ satisfies $L_2 g = 0$ with $L_2 \equiv (d^2/dt^2) + \bar{\omega}^2$ (see Example (g)). This means that

$$\left(\frac{d^2}{dt^2} + \bar{\omega}^2\right)\left(\frac{d^2}{dt^2} + 2b\frac{d}{dt} + \omega^2\right)Q = 0.$$

The associated characteristic equation is $(m^2 + \bar{\omega})(m^2 + 2bm + \omega^2) = 0$, and here we shall consider only the case when $b < \omega$. Since the values of m coming from $m^2 + 2bm + \omega^2 = 0$ are already contained in the solution to the original homogeneous equation, we need only $Q_p \equiv \alpha_1 \cos\bar{\omega}t + \alpha_2 \sin\bar{\omega}t$. The constants α_1, α_2 will be determined by substituting Q_p in the original inhomogeneous equation

$$\alpha_1(-\bar{\omega}^2\cos\bar{\omega}t - 2b\bar{\omega}\sin\bar{\omega}t + \omega^2\cos\bar{\omega}t) + \alpha_2(-\bar{\omega}^2\sin\bar{\omega}t + 2b\bar{\omega}\cos\bar{\omega}t + \omega^2\sin\bar{\omega}t)$$
$$= e_0 \sin\bar{\omega}t.$$

Equating coefficients of $\cos\bar{\omega}t$ and $\sin\bar{\omega}t$ on both sides, we get $\alpha_1(\omega^2 - \bar{\omega}^2) + 2b\bar{\omega}\alpha_2 = 0$ and $\alpha_2(\omega^2 - \bar{\omega}^2) - 2b\bar{\omega}\alpha_1 = e_0$ which we can solve to yield α_1 and α_2. This can then be substituted back to get a particular integral Q_p, which on addition with the general solution Q_h of the homogeneous equation, provides the general solution of the total inhomogeneous problem.

4.7 Some other applications of Calculus

In this section we shall consider a few applications of calculus such as the concept of maxima–minima or optimization, and fixed point theorems which do not involve differential equations.

Maxima–Minima

Given a smooth (i.e., arbitrarily often differentiable) function f of a real variable, how do we find where f is increasing and where it is decreasing? For example, consider $f(x) = x^3 - 3x^2 - 9x + 1$. It can be shown that the region where f is increasing will be decided by where $f'(x) \geq 0$ (similarly f decreasing by $f'(x) \leq 0$) (**Exercise:** for proving that $f' \geq 0$ implies f is increasing, use the fundamental theorem of integral calculus). For this, we first find the point where the derivative of f vanishes, i.e., $f'(x) = 0$. In this case, we get $f'(x) = 3(x^2-2x-3) = 3(x+1)(x-3)$ and this vanishes only for $x = -1$ and 3. It also follows from the above expression for f' that $f'(x) > 0$ for $x < -1$ and for $x > 3$, while $f'(x) < 0$ if $-1 < x < 3$. Thus the graph of f looks as in Figure 4.13. We see that the points x for which $f'(x) = 0$ (or undefined) play a crucial role in deciding what the graph of f looks like and are called the *critical points* of the function f, while the values of f at the critical points are called the *critical values* of f.

By looking at the graph, we can also see that geometrically the critical points of f are precisely those where the tangent to the graph of f is parallel to the x-axis. At a critical point where f' is not defined at all, either the tangent is perpendicular to the x-axis (e.g., for $f(x) = \sqrt{x}$ for $x \geq 0$ and $= -\sqrt{-x}$ for $x \leq 0$ at $x = 0$) or there is no tangent at all (e.g., for $f(x) = |x|$ at $x = 0$). Here we shall be concerned only with smooth functions and therefore our critical points will be solutions of $f'(x) = 0$. It is clear from the above discussion that the critical points divide the domain of definition of f into intervals on which the sign of the derivative stays the same and therefore, between two successive critical points the graph of a function either goes up or down. Of course, a function may not have any critical point (e.g., $f(x) = x$) or may have infinitely many of them (e.g., $f(x) = \sin x$).

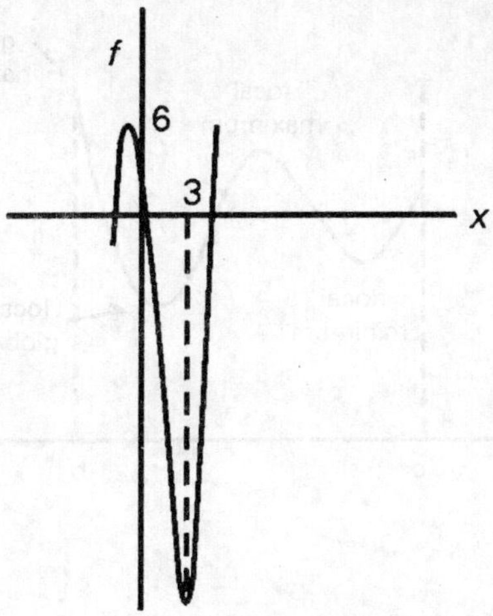

Figure 4.13

We can discern another phenomenon by looking at the graph in Figure 4.13 viz., there is a change of direction in the graph precisely at the critical points. We see that f has a *local maximum* at $x = -1$ since near $x = -1$, the values of f get no larger than $f(-1)$ and also that f has a *local minimum* at $x = 3$ since the values of f get no smaller than $f(3)$ near $x = 3$. Thus if $x = x_0$ is a critical point of a function f and if f' changes sign at x_0, then either (a) f' is positive to the left of x_0 and negative to its right, in which case f has a local maximum at x_0 or, (b) it is just the reverse of (a), in which case f has a local minimum at x_0. This can be re-expressed as: if $f'(x_0) = 0$ and f' changes sign at x_0, then either (a) $f''(x_0) < 0$ for a local maximum at x_0 or (b) $f''(x_0) > 0$ for a local minimum at x_0. In the example above, $f''(x) = 6(x-1)$, $f''(-1) = -12$ and $f''(3) = 12$, confirming our conclusion. Not every solution of $f'(x) = 0$ corresponds to a local maxima or local minima: consider $f(x) = x^3$ at $x = 0$!

So far we have discussed critical points in the neighbourhood of which the value of a function is smaller or larger than the critical value. However, sometimes we may be interested in where the function is absolutely largest or smallest in the given domain or equivalently, in a *global maximum* or *global minimum* of a function. For example, consider the graph in Figure 4.14.

Some caution is necessary in dealing with global maximum or minimum: *not every global maximum or minimum is necessarily a critical point* (e.g., $f(x) = x$ for $0 \le x \le 1$ has global maximum at $x = 1$ but $f'(x) = 1$ for all x).

In analogy with the classification of mirrors and lenses in optics, we say that the function is *convex* if its graph bends upwards and is *concave* if its graph bends downwards. More precisely, f is convex if the value of f at any point x_0 in a line segment joining any two points a, b in its domain is less than or equal to the segment joining the values of f at a and b in the same ratio with which x_0 occurs in $[a, b]$, i.e.,

$$f(x_0) \le f(a)\frac{b - x_0}{b - a} + f(b)\frac{x_0 - a}{b - a}.$$

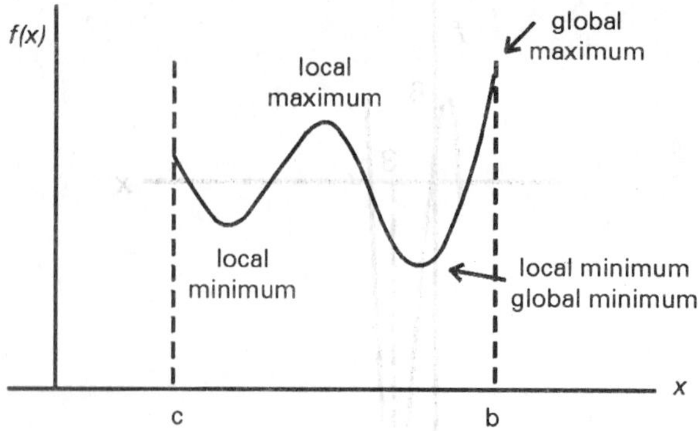

Figure 4.14

Similarly f is concave if "less than or equal to" in the above sentence is replaced by "greater than or equal to". Note however, a straight line is neither convex nor concave. The next figure (Figure 4.15) will help explain this.

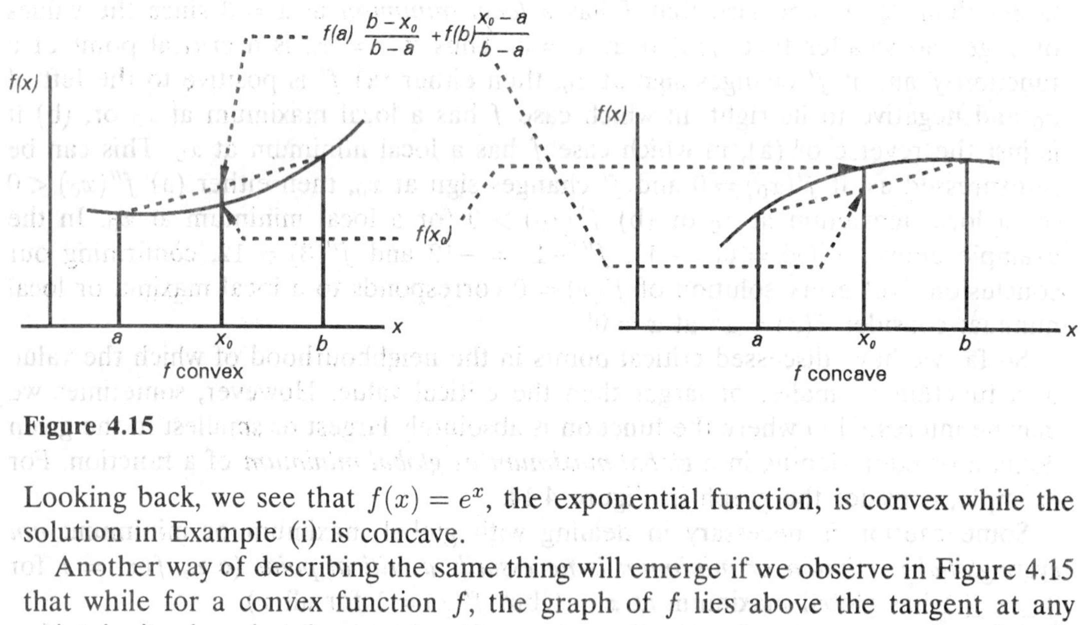

Figure 4.15

Looking back, we see that $f(x) = e^x$, the exponential function, is convex while the solution in Example (i) is concave.

Another way of describing the same thing will emerge if we observe in Figure 4.15 that while for a convex function f, the graph of f lies above the tangent at any point in its domain of convexity, the converse is true for a concave function. It is also clear from the figure that if f is smooth, then f is convex near x_0 if and only if $f''(x) \geq 0$ near x_0, and f is concave near x_0 if and only if $f''(x) \leq 0$ near x_0. (**Exercise:** Prove this statement from the earlier definition of a convex or concave function). Thus if x_0 is a local maximum or minimum for f, then f is concave or convex respectively near x_0. The converse need not be true since a function f (as in Figure 4.15) may be concave or convex near x_0 without x_0 being a critical point of f.

Finally, the point x_0 at which the function changes from convexity to concavity or the other way around is called the *point of inflection* of f, and is given by the solution of the equation $f''(x_0) = 0$ and $f'''(x) \neq 0$. With reference to Figure 4.13, we see that the point $x = 1$ is a point of inflection.

Examples

(a) In our economic model we shall assume for simplicity that the decisions in a firm or industry depend on the cost of production of some goods, say cars and the revenue or income accrued from their sale. The *cost function* $C(q)$ gives the total cost of production of a quantity q of cars, which for our simple model shall be assumed to take all positive real values though in reality it should take only integer values.

The graph of C usually looks as in Figure 4.16: the intercept on the C-axis is the *fixed cost* which is incurred even if nothing is produced, and initially the cost function increases quickly before settling down to a somewhat slower growth because producing larger quantities of cars is more efficient than producing smaller quantities. On the other hand, the revenue function, $R(q)$, represents the total income received by the firm from the sale of q cars. Here again initially R is linear, but for larger values of q, the market may get saturated with cars, causing the price and hence revenue to drop leading to the following shape of the graph of R (Figure 4.17). Often various decisions for the firm are dictated by the profit, designated by P, given by $P(q) = R(q) - C(q)$, and the profit curve looks as in Figure 4.18.

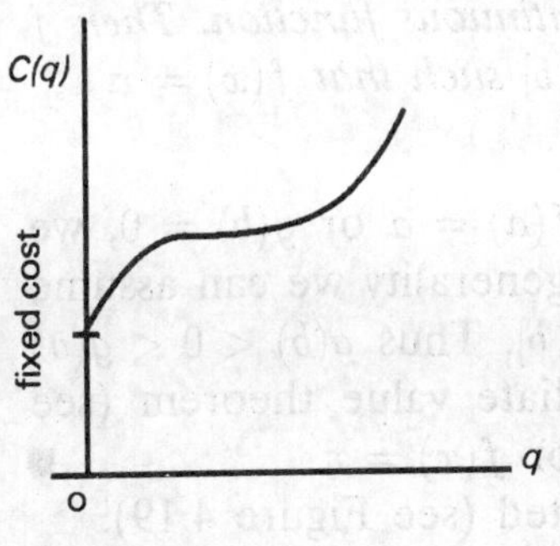

Figure 4.16

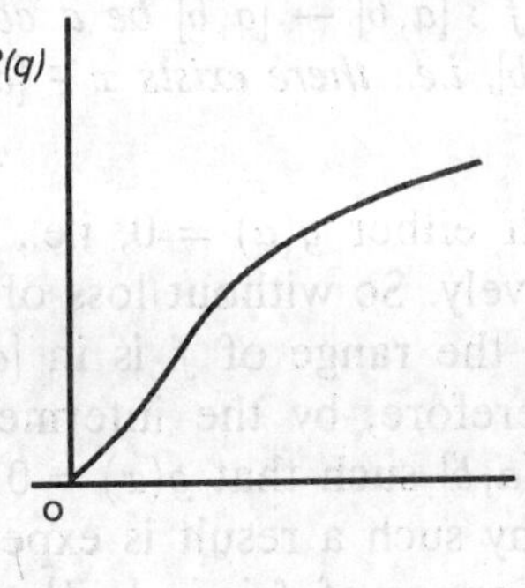

Figure 4.17

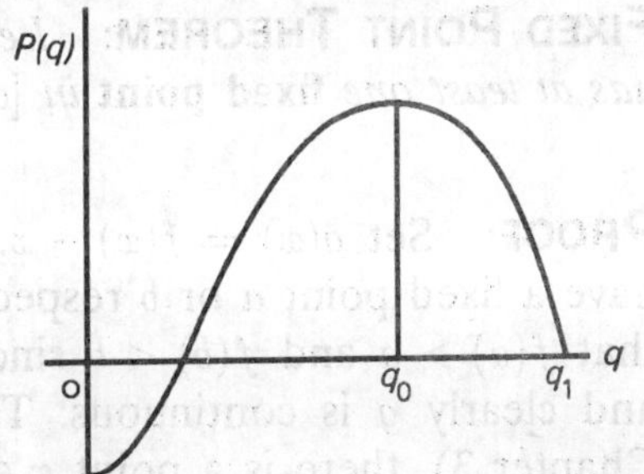

Figure 4.18

In fact the graph of P is obtained by looking at the vertical differences between the graphs of R and C. If the firm wants to maximize its profit, we have to look for the solution of $P'(q) = 0$ or $R'(q) = C'(q)$, i.e., look for the critical points of P. The functions C' and R' are called the marginal cost and marginal revenue functions respectively and measure approximately the additional cost incurred (or revenue generated) per unit quantity. From the graph of P in Figure 4.18, it is clear that global maximum is achieved for $q = q_0$, while minimum occurs at $q = 0$ and $q = q_1$. Thus maximum or minimum of profit is attained when the marginal cost equals marginal revenue.

(b) We want to design a metallic cylindrical can such that it encloses a given volume, say 200 ml, of a soft drink while the surface area is minimum so that the cost of production of the can is minimum. The volume of the can is $\pi r^2 h$, where r is the radius and h is the height of the can, and the surface area $S = 2\pi r^2 + 2\pi rh$. Thus we will have to minimize S, holding $\pi r^2 h = 200$. Eliminating h from the two expressions, we get

$$S(r) = 2\pi r^2 + \frac{400}{r}$$

and $S'(R) = 4\pi r - (400/r^2) = 0$ yields one solution $r = (100/\pi)^{1/3} = 3.17$ cm. In this case it is also clear that this is a global minimum (**Exercise!**).

Fixed point theorem

Here we discuss two elementary, but rather deep, consequences of the intermediate value theorem proven in Chapter 3.

FIXED POINT THEOREM: *Let $f : [a, b] \to [a, b]$ be a continuous function. Then f has at least one* **fixed point** *in $[a, b]$, i.e., there exists $x \in [a, b]$ such that $f(x) = x$.*

PROOF: Set $g(x) = f(x) - x$. If either $g(a) = 0$, i.e., $f(a) = a$ or $g(b) = 0$, we have a fixed point a or b respectively. So without loss of generality we can assume that $f(a) > a$ and $f(b) < b$ since the range of f is in $[a, b]$. Thus $g(b) < 0 < g(a)$ and clearly g is continuous. Therefore, by the intermediate value theorem (see Chapter 3), there is a point $x \in [a, b]$ such that $g(x) = 0$ or $f(x) = x$. ∎

Graphically it is easy to see why such a result is expected (see Figure 4.19).

Since both the domain and the range of f is in $[a, b]$, we need to look inside the square $[a, b] \times [a, b]$ and look for the points of intersection of the graphs of $y = x$ and $y = f(x)$. In Figure 4.19 there are 3 fixed points x_1, x_2 and x_3.

The above theorem is a special case of Brouwer's fixed point theorem which asserts that every continuous function $f : B^d \to B^d$ (the closed unit ball in d-dimensions) has at least one fixed point (see for example, reference [3], Chapter 7). There are other fixed point theorems also and they have extensive applications in many areas.

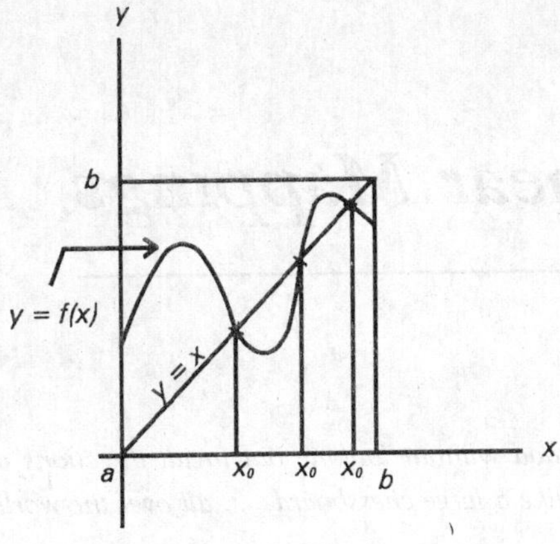

Figure 4.19

We shall end this chapter with a brief discussion of another remarkable theorem, Sarkovskii's theorem, whose proof depends only on the intermediate value theorem. Let $f : \mathbb{R} \to \mathbb{R}$ be continuous, then x is said to be a *periodic point* of f if there is a natural number n such that $f^n(x) \equiv f \circ f \circ \cdots \circ f(x)(n - \text{times}) = x$. The least such n is called the prime period for the point x. Then we have a simplified version:

SARKOVSKII'S THEOREM: *Let $f : \mathbb{R} \to \mathbb{R}$ be continuous and assume furthermore that f has a periodic point of period 3. Then f has periodic points of all other periods.*

We do not give a proof of this which, though elementary, is long and the reader is referred to the book [4] for the proof and many other interesting observations in one and two-dimensional dynamical systems theory.

Suggested Reading

1. *Calculus*, Vol. I by T.M. Apostol, Wiley Eastern, New Delhi 1980 (read Chapter 8).

2. *Calculus with Applications and Computing* by P. Lax, S. Burstein and A. Lax, Narosa Publishing House, New Delhi 1979 (read Chapters 5, 8 and 9).

3. *From Calculus to Cohomology and Characteristic classes* by I.Madsen and J.Tornehave, Cambridge University Press 1997 (read Chapter 1).

4. *An Introduction to Chaotic Dynamical Systems* by R.L.Devaney, Addison Wesley, New York 1988. (read Chapter 1).

Chapter 5

Linear Mappings

"For some minutes Alice stood without looking out in all directions over the country . . . I declare it's marked out just like a large chessboard . . . all over the world—if this is the world at all."

Lewis Carroll

5.1 Introduction

We are familiar with the notion of a mapping or function from Chapter 1, where we encountered several examples. We also learnt that with each mapping is associated a domain and a range. In this chapter, we shall be interested in a special set of mappings. For example, consider the mapping from $\mathbb{N}$ into itself given by the expression $f(x) = 2x$. One observes immediately that it satisfies the additive property $f(x + y) = f(x) + f(y)$.

Similarly, recall the equation of a straight line passing through the origin: $y = mx$, where m is the slope of the straight line. It is clear that this expression represents the mapping f from the set of real numbers $\mathbb{R}$ into itself: $f(x) = mx$, and that it enjoys the properties

$$f(x + y) = f(x) + f(y); \ f(\alpha x) = \alpha f(x) \tag{5.1}$$

for all real numbers α, x and y.

The property (5.1) of the function f appears in many other situations, e.g., in Newton's second law of motion for a particle in motion along a straight line: $F = f(a) = ma$, where m is the mass of the particle, a is the acceleration being experienced by the particle and F is the force experienced by the particle. We can easily check that this mapping f satisfies the two properties given in (5.1).

Any mapping f from the set of real numbers into itself satisfying the properties listed in (5.1) is called a *linear mapping.*

Exercise: Show that any continuous mapping f from the set of real numbers into the set of real numbers which satisfies only the first property in (5.1), must necessarily have the form $f(x) = mx$, for some fixed real number m and for all real numbers x. (Hint: Use the fact from Chapter 2 that any real number is the limit of a sequence of rational numbers).

5.2 Linear Mappings on $\mathbb{R}^2$

We now consider linear mappings on $\mathbb{R}^2$, i.e., on the points in a plane. The moment we fix our coordinate axes in the Cartesian plane we can represent each point in the plane by means of an ordered pair of real numbers, say (x, y). The plane itself is represented by the symbol $\mathbb{R}^2$ which is shorthand for $\mathbb{R} \times \mathbb{R}$, the Cartesian product of the set of real numbers $\mathbb{R}$ with itself. There are certain very natural properties that $\mathbb{R}^2$ inherits from $\mathbb{R}$, some of which we outline below. To begin with, we define the following two algebraic operations on $\mathbb{R}^2$.

Addition and Scalar Multiplication in $\mathbb{R}^2$

For all elements (a, b) and (c, d) in $\mathbb{R}^2$ and $\alpha \in \mathbb{R}$,

$$(a, b) + (c, d) = (a + c, b + d) \quad \text{(addition)},$$

$$\alpha(a, b) = (\alpha a, \alpha b) \quad \text{(scalar multiplication)}.$$

We must not fail to observe that even though we have retained the same symbol $+$ on both sides of the equation (5.2), it has two different connotations. On the left side of (5.2) it represents a combination of two elements of $\mathbb{R}^2$. This way of combining, also called addition, is defined by the expression on the right hand side of (5.2). If we recall the idea of representing a vector by a point in a plane then we will at once notice that addition and multiplication that we have described in (5.2) and (5.3) are nothing but the parallelogram law of addition and the dilation respectively of the vectors. In fact $\mathbb{R}^2$ equipped with the two operations as defined by (5.2) and (5.3) is called a *vector space* over $\mathbb{R}$ and enjoys certain formal properties as listed below.

Properties of Addition in $\mathbb{R}^2$

For all elements $(a, b), (c, d)$ and (x, y) of $\mathbb{R}^2$:

$A(i)$ $(a, b) + (c, d) = (c, d) + (a, b)$, (commutativity),

$A(ii)$ $((a, b) + (c, d)) + (x, y) = (a, b) + ((c, d) + (x, y))$, (associativity),

$A(iii)$ $(a, b) + (0, 0) = (a, b)$, (additive identity),

$A(iv)$ $(a, b) + (-a, -b) = (0, 0)$, (additive inverse).

Properties of Scalar Multiplication in $\mathbb{R}^2$

For all real numbers α, β and for all elements $(a, b), (c, d)$ of $\mathbb{R}^2$:

$M(i)$ $\alpha((a, b) + (c, d)) = \alpha(a, b) + \alpha(c, d)$, (distributivity),

$M(ii)$ $(\alpha + \beta)(a, b) = \alpha(a, b) + \beta(a, b)$, (distributivity),

$M(iii)$ $(\alpha(\beta(a, b)) = (\alpha\beta)(a, b)$, (associativity),

$M(iv)$ $0(a, b) = (0, 0)$, (5.2)

$M(v)$ $\alpha(0, 0) = (0, 0)$. (5.3)

We shall denote elements of $\mathbb{R}^2$ by boldface letters $\boldsymbol{x}, \boldsymbol{y}, \boldsymbol{z}$, etc. Hence if the letter $\boldsymbol{x}$ stands for the element (a, b) of $\mathbb{R}^2$ and α is a real number, then the meaning of $\alpha\boldsymbol{x}$ should be the element $(\alpha a, \alpha b)$ of $\mathbb{R}^2$, i.e. $\alpha\boldsymbol{x} = (\alpha a, \alpha b)$. Similarly, when we say that $\boldsymbol{x}$ and $\boldsymbol{y}$ are two elements of $\mathbb{R}^2$ say (a, b) and (c, d) respectively, then $\boldsymbol{x} + \boldsymbol{y}$ stands for that element of $\mathbb{R}^2$ which is the element $(a + c, b + d)$ of $\mathbb{R}^2$. Consider the mapping T from $\mathbb{R}^2$ into $\mathbb{R}^2$ given by the rule

$$T(\boldsymbol{x}) = 2\boldsymbol{x} \tag{5.4}$$

for each element $\boldsymbol{x}$ of $\mathbb{R}^2$. So if (a, b) is a typical element of $\mathbb{R}^2$, then T is that mapping which sends (a, b) to the element $2(a, b)$ which is equal to the element $(2a, 2b)$ of $\mathbb{R}^2$. It follows that

$$T(\boldsymbol{x} + \boldsymbol{y}) = T(\boldsymbol{x}) + T(\boldsymbol{y}), \tag{5.5}$$

$$T(\alpha\boldsymbol{x}) = \alpha T(\boldsymbol{x}) \tag{5.6}$$

for all x, y in $\mathbb{R}^2$ and for all α in $\mathbb{R}$.

Exercise: Verify (5.5) and (5.6) for the example in (5.4) and show that in general, each of them implies $T(0,0) = (0,0)$.

DEFINITION: Any mapping T from $\mathbb{R}^2$ which satisfies (5.5) and (5.6) is called a *linear mapping* or a *linear transformation* on $\mathbb{R}^2$.

As seen above, (5.4) is an example of a linear mapping (linear transformation) from $\mathbb{R}^2$ to $\mathbb{R}^2$. However, every linear transformation on $\mathbb{R}^2$ need not be of the form $T(x) = \alpha x$ for some fixed real number α. For example, consider the mapping T on $\mathbb{R}^2$ given by

$$T(a,b) = (b,a) \quad \text{for all} \quad (a,b) \quad \text{in} \quad \mathbb{R}^2. \tag{5.7}$$

It is easy to verify that T given by (5.7) is a linear mapping on $\mathbb{R}^2$ but that there does not exist a real number α so that $T(x) = \alpha x$ for all x. The same linear mapping T has the geometrical interpretation: that it is the reflection about the line $y = x$ in $\mathbb{R}^2$.

Exercises

Check which of the following mappings are linear. Which of these mappings can be given a geometrical interpretation? The mappings are defined for all (a,b) in $\mathbb{R}^2$.

(i) $T(a,b) = (2a, 3b)$,

(ii) $T(a,b) = (-a, b)$,

(iii) $T(a,b) = (a,b)$,

(iv) $T(a,b) = (a,0)$,

(v) $T(a,b) = (ab, b)$,

(vi) $T(a,b) = 3(a; (a+b))$,

(vii) $T(a,b) = (a,a)$,

(viii) $T(a,b) = (a+b, a-b)$,

(ix) $T(a,b) = (\alpha a + \beta b, \gamma a + \delta b)$,

where α, β, γ and δ are arbitrarily chosen and fixed real numbers.

Exercise: Having identified the linear mappings from the above list, show that each of these mappings is actually a special case of Exercise (ix).

Next we shall show that (ix) is the most general form of a linear mapping T on $\mathbb{R}^2$. By the properties $A(i)$ and $M(i)$, we see that an element (a,b) of $\mathbb{R}^2$ can be written as

$$(a,b) = a(1,0) + b(0,1). \tag{5.8}$$

Set $T(1,0) = (\alpha,\gamma)$ and $T(0,1) = (\beta,\delta)$. Then by (5.5) and (5.6),

$$\begin{aligned} T(a,b) &= T(a(1,0) + b(0,1)) \\ &= T(a(1,0)) + T(b(0,1)) \\ &= aT(1,0) + bT(0,1) \\ &= a(\alpha,\gamma) + b(\beta,\delta) \\ &= (\alpha a, \gamma a) + (\beta b, \delta b) \\ &= (\alpha a + \beta b, \gamma a + \delta b), \end{aligned}$$

so that T is completely determined by the numbers α, β, γ and δ.

5.3 *The Algebra of Linear Transformations on* $\mathbb{R}^2$

As in $\mathbb{R}^2$, we shall see that the set of linear mappings on $\mathbb{R}^2$ (written $\mathcal{L}(\mathbb{R}^2)$ for short) themselves satisfy the properties A and M. For this we need to define the operations of addition and scalar multiplication in $\mathcal{L}(\mathbb{R}^2)$ as follows. For S and T in $\mathcal{L}(\mathbb{R}^2)$, the sum of S and T is the mapping $S+T$ given by

$$(S+T)(x) = S(x) + T(x). \tag{5.9}$$

Similarly for α in $\mathbb{R}$, αT is defined by

$$(\alpha T)(x) = \alpha T(x) \tag{5.10}$$

for all x in $\mathbb{R}^2$.

There is a new set of operations in the class $\mathcal{L}(\mathbb{R}^2)$, namely that of *product* or *multiplication* of two linear mappings. For S and T in $\mathcal{L}(\mathbb{R}^2)$, we define ST by

$$ST(\boldsymbol{x}) = S \circ T(\boldsymbol{x}) \text{ for } \boldsymbol{x} \in \mathbb{R}^2_1, \tag{5.11}$$

where $\circ$ stands for composition of functions as in Section 1.9.

Exercise: Show that the maps $S+T, \alpha T$ and ST, defined in (5.9)–(5.11), are linear.

The set $\mathcal{L}(\mathbb{R}^2)$ equipped with the three operations defined above satisfies the following properties.

For all real numbers α and β and for all elements S, T and F of $\mathcal{L}(\mathbb{R}^2)$,

A(i) $S+(T+F) = (S+T)+F$ and $(ST)F = S(TF)$.

A(ii) $S+T = T+S$.

A(iii) $(S+T)F = SF+TF$ and $F(S+T) = FS+FT$.

A(iv) $(\alpha+\beta)S = \alpha S + \beta S$.

A(v) $(\alpha S)T = \alpha(ST)$.

A(vi) There exists a unique element I in $\mathcal{L}(\mathbb{R}^2)$ such that $SI = IS = S$, for all S in $\mathcal{L}(\mathbb{R}^2)$.

A(vii) There exists a unique element O in $\mathcal{L}(\mathbb{R}^2)$ such that $T+O = O+T = T$, for all T.

A(viii) For each T there exists a unique S in $\mathcal{L}(\mathbb{R}^2)$ such that $S+T = T+S = O$.

(It can be shown from A(vii) using A(viii) that $TO = OT = O$ for all T.)

These properties are rather easy to prove and we shall leave the proofs as an **Exercise** (Hint: take $I(\boldsymbol{x}) = \boldsymbol{x}$ and $O(\boldsymbol{x}) = 0$ for all $\boldsymbol{x} \in \mathbb{R}^2$). Thus $\mathcal{L}(\mathbb{R}^2)$ is an algebra.

We have seen how $\mathcal{L}(\mathbb{R}^2)$ has operations that resemble properties of operations on numbers. However, there are some significant differences and we highlight two of them below.

D(i) There exist elements S, T in $\mathcal{L}(\mathbb{R}^2)$ such that $ST \neq TS$.

Choose S as $S(a,b) = (a+b,0)$ and T as $T(a,b) = (b,a)$ for all (a,b) in $\mathbb{R}^2$. Then $ST(a,b) = (a+b,0)$ and $TS(a,b) = (0,a+b)$ and hence there are an infinite number of elements (a,b) of $\mathbb{R}^2$ for which $ST(a,b) \neq TS(a,b)$. Thus $ST \neq TS$.

D(ii) There exist non-zero elements S and T of $\mathcal{L}(\mathbb{R}^2)$ such that $ST = O$.

This is also easy to see. Let S be given by $S(a,b) = (a,0)$ and let T be given by $T(a,b) = (0,a+b)$. Clearly S and T are non-zero but ST is O, i.e., ST is that linear transformation which takes each element to the element $(0,0)$.

5.4 Linear Transformations and Matrices

A finite collection of elements $\{\boldsymbol{x}_1, \ldots, \boldsymbol{x}_k\}$ of $\mathbb{R}^2$ is said to be linearly dependent if we can find real numbers $\{\alpha_1, \ldots, \alpha_k\}$ with at least one α_i different from zero such that

$$\sum_{j=1}^{k} \alpha_j \boldsymbol{x}_j = \mathbf{0}.$$

If a set $\{\boldsymbol{x}_1, \ldots, \boldsymbol{x}_k\}$ is not linearly dependent then it is said to be *linearly independent*. This simply means that if

$$\sum_{j=1}^{k} \alpha_j \boldsymbol{x}_j = \mathbf{0},$$

then each α_i is zero. Clearly, $\mathbf{0}$ can never be a member of a linearly independent set. (Note: any expression of the form $\sum_{j=1}^{k} \alpha_i \boldsymbol{x}_j$ is said to be a linear combination of the elements $\boldsymbol{x}_1, \ldots, \boldsymbol{x}_k$.).

The set $\{(1,0),(0,1)\}$ can be seen to be a linearly independent set. The set $\{(1,0),(0,1)\}$ also has another nice property, namely it spans $\mathbb{R}^2$, i.e., each element of $\mathbb{R}^2$ is a linear combination of $(1,0)$ and $(0,1)$. Any linearly independent set which spans $\mathbb{R}^2$ is called a *basis* of $\mathbb{R}^2$ so that $\{(1,0),(0,1)\}$ is a basis of $\mathbb{R}^2$. It is called the standard basis of $\mathbb{R}^2$. Here is another example of a basis in $\mathbb{R}^2$: $\{(1,1),(0,1\}$. To see that this set is a basis is not difficult. First we must check its linear independence. Suppose there are two scalars α and β such that $\alpha(1,1) + \beta(0,1) = \mathbf{0}$. This means that $(\alpha, \alpha+\beta) = (0,0)$. This forces us to conclude that $\alpha = 0$ and $\alpha + \beta = 0$. Thus both the scalars are zero, which establishes the linear independence of the set. To show that this set spans $\mathbb{R}^2$ is just as easy. Let (a,b) be a typical element of $\mathbb{R}^2$. Then clearly we have $(a,b) = t(0,1) + s(1,1)$, where $t = b - a$ and $s = a$.

Exercise: Let $x, y \in \mathbb{R}^2$. Suppose

$$x = (a, b) \text{ and } y = (c, d).$$

(i) Show that $\{x, y\}$ is linearly independent if and only if

$$ad - bc \neq 0.$$

(ii) Suppose $\{x, y\}$ is linearly independent. Let $z = (u, v) \in \mathbb{R}^2$. Show that there exists unique $\alpha, \beta \in \mathbb{R}$ such that

$$z = \alpha x + \beta y.$$

[**Hint**: Solve simultaneous equations and use (i)].

(iii) Let $x, y, z \in \mathbb{R}^2$ and be distinct. Show that $\{x, y, z\}$ cannot be linearly independent.

(iv) Show that any two linearly independent elements of $\mathbb{R}^2$ form a basis.

In fact, the converse is also true, i.e.,

THEOREM: *Any basis of $\mathbb{R}^2$ has exactly two elements.*

The proof of the theorem is left as an **Exercise** (use (iii) above and show that one element in $\mathbb{R}^2$ cannot span $\mathbb{R}^2$).

One can restate the above discussion by saying that there is greater room in $\mathbb{R}^2$ compared to $\mathbb{R}$. If two elements of $\mathbb{R}^2$ are not linearly independent, then obviously one lies on the straight line joining the other element to the origin and extended indefinitely in both directions.

Exercise: Give examples to show that two maps that are not linear can agree on $(0, 1)$ and $(1, 0)$ and still be different.

Recall the notion of the inverse of a function from Section 1.10. In the context of a linear transformation on $\mathbb{R}^2$, the same ideas are applicable for the condition of invertibility of a linear transformation as a function. We can think of a linear transformation T on $\mathbb{R}^2$ as a function from $\mathbb{R}^2$ to $\mathbb{R}^2$. Suppose furthermore that T is bijective (i.e., one-to-one and onto), then T will have an inverse, denoted by T^{-1}. Next, let $x, y \in \mathbb{R}^2$and $\alpha \in \mathbb{R}$. Since T is onto, there are elements u and v in $\mathbb{R}^2$ such that $T(u) = x$ and $T(v) = y$.

Hence

$$\begin{aligned}
T^{-1}(x+y) &= T^{-1}(T(u)+T(v)) \\
&= T^{-1}(T(u+v)) \quad (\text{ as } T \text{ is a linear transformation}) \\
&= T^{-1}T(u+v) \\
&= u+v \quad (\text{because } T^{-1}T = \text{Identity}) \\
&= T^{-1}Tu + T^{-1}Tv \quad \text{because } T^{-1}T = \text{Identity}) \\
&= T^{-1}x + T^{-1}y \quad (\text{as } Tu = x \text{ and } Tv = y).
\end{aligned}$$

Similarly, one can show that $T^{-1}(\alpha x) = \alpha T^{-1}(x)$, leading to:

THEOREM: *The inverse of a linear transformation, if it exists, is also a linear transformation.*

Exercise: Show that a linear transformation T is one-to-one if and only if $Tx = 0$ implies $x = 0$.

In fact, the assumption of T being bijective is not necessary since one of the properties implies the other as is shown in the next theorem.

THEOREM: *Let T be a linear transformation on $\mathbb{R}^2$. Then T is one-to-one if and only if it is onto.*

PROOF: By Exercise (ix) on page 145, $T(a,b) = (\alpha a+\beta b, \gamma a+\delta b)$. By the preceding exercise, T is one-to-one if and only if $T(a,b) = (0,0)$ implies $a = b = 0$, or equivalently $\alpha a + \beta b = \gamma a + \delta b = 0$ implies $a = b = 0$. Clearly, this happens only if $\alpha\delta - \beta\gamma \neq 0$. Let $z = (u,v) \in \mathbb{R}^2$, then T is onto if and only if the equations $\alpha a + \beta b = u$ and $\gamma a + \delta b = v$ have the unique solution (u,v). As we have seen in Exercise (ii) on page 149, the condition $\alpha\delta - \beta\gamma \neq 0$ precisely ensures this. Thus T being one-to-one implies T is onto. The converse implication follows by reversing the steps. ∎

Exercise: Show that if S and T are linear transformations on $\mathbb{R}^2$, then ST is invertible if and only if both S and T are invertible. In that case express the inverse of ST in terms of the inverses of S and T.

Remark: The proof of the above theorem shows that T given by $T(a,b) = (\alpha a + \beta b, \gamma a + \delta b)$ is invertible if and only if $\alpha\delta - \beta\gamma \neq 0$.

We can see from the above discussion that to each linear transformation T on $\mathbb{R}^2$ we associate four scalars α, β, γ and δ as follows:

$$T(1,0) = \alpha(1,0) + \gamma(0,1),$$
$$T(0,1) = \beta(1,0) + \delta(0,1). \tag{5.12}$$

Given these four numbers we form an array consisting of two rows and two columns in the following way:

$$\begin{bmatrix} \alpha & \beta \\ \gamma & \delta \end{bmatrix}. \tag{5.13}$$

This array, called a 2×2 *matrix*, corresponds to the transformation T with respect to the basis $\{(1,0),(0,1)\}$, and is denoted by M_T. Also, given a 2×2 matrix in (5.13) above, we can define a linear transformation T on $\mathbb{R}^2$ as follows:

$$T(a,b) = (\alpha a + \beta b, \gamma a + \delta b). \tag{5.14}$$

Exercise: Show that the correspondence $M : \mathcal{L}(\mathbb{R}^2) \to 2 \times 2$ matrices with real entries given by $M(T) = M_T$ is one-to-one and onto.

Remark: The positions of the numbers in (5.13) are fixed and they are determined by the order of their occurrence in (5.13), but note that the order of occurrence in (5.12) is different from the order of occurrence in (5.13). If we designate the entries of the matrix M_T by $t_{ij}(i,j = 1,2)$ (i is the row-index while j is that for the columns), then in (5.13), $t_{11} = \alpha, t_{12} = \beta, t_{21} = \gamma, t_{22} = \delta$.

5.5 *Algebra of Matrices*

We have introduced the algebraic properties of $\mathcal{L}(\mathbb{R}^2)$ in Section 5.3 and have seen in the previous section the correspondence between linear transformations in $\mathbb{R}^2$ and 2×2 matrices. Here we transcribe the algebraic properties of $\mathcal{L}(\mathbb{R}^2)$ into those for the set of 2×2 matrices.

Given $S, T \in \mathcal{L}(\mathbb{R}^2)$ and $\theta \in \mathbb{R}$, we know from Section 5.3 that $S + T, \theta T, ST \in \mathcal{L}(\mathbb{R}^2)$. On the other hand, the exercise in the last section tells us that the association $T \to M_T$ is one-to-one and onto. Therefore a natural question arises: can we define addition, scalar multiplication and product for 2×2 matrices such that

$$M_{T+S} = M_T + M_S,$$

$$M_{\theta T} = \theta M_T,$$

$$M_{ST} = M_S M_T \ ? \tag{5.15}$$

Now, let the matrices corresponding to S and T be

$$M_S = \begin{bmatrix} \alpha & \beta \\ \gamma & \delta \end{bmatrix} \text{ and } M_T = \begin{bmatrix} \alpha' & \beta' \\ \gamma' & \delta' \end{bmatrix}.$$

The first two properties in (5.15) are easily verified provided we define the addition and scalar multiplication for 2×2 matrices as

$$\begin{bmatrix} \alpha & \beta \\ \gamma & \delta \end{bmatrix} + \begin{bmatrix} \alpha' & \beta' \\ \gamma' & \delta' \end{bmatrix} = \begin{bmatrix} \alpha+\alpha' & \beta+\beta' \\ \gamma+\gamma' & \delta+\delta' \end{bmatrix} \text{ and}$$

$$\theta \begin{bmatrix} \alpha' & \beta' \\ \gamma' & \delta' \end{bmatrix} = \begin{bmatrix} \theta\alpha' & \theta\beta' \\ \theta\gamma' & \theta\delta' \end{bmatrix}.$$

For the last property in (5.15), we note from (5.12) that

$$S(1,0) = \alpha(1,0) + \gamma(0,1), \ \ S(0,1) = \beta(1,0) + \delta(0,1),$$

$$T(1,0) = \alpha'(1,0) + \gamma'(0,1), \ \ T(0,1) = \beta'(1,0) + \delta'(0,1). \tag{5.16}$$

Therefore, by (5.16) and the linearity of S,

$$\begin{aligned} (ST)(1,0) &= S\{\alpha'(1,0) + \gamma'(0,1)\} = \alpha' S(1,0) + \gamma' S(0,1) \\ &= \alpha'\{\alpha(1,0) + \gamma(0,1)\} + \gamma'\{\beta(1,0) + \delta(0,1)\} \\ &= (\alpha\alpha' + \beta\gamma')(1,0) + (\gamma\alpha' + \delta\gamma')(0,1). \end{aligned}$$

Similarly, one finds that

$$(ST)(0,1) = (\alpha\beta' + \beta\delta')(1,0) + (\gamma\beta' + \delta\delta')(0,1).$$

Thus the last property of (5.15) is indeed verified if we define the product of the matrices M_S and M_T as

$$\begin{bmatrix} \alpha & \beta \\ \gamma & \delta \end{bmatrix} \begin{bmatrix} \alpha' & \beta' \\ \gamma' & \delta' \end{bmatrix} = \begin{bmatrix} \alpha\alpha' + \beta\gamma' & \alpha\beta' + \beta\delta' \\ \gamma\alpha' + \delta\gamma' & \gamma\beta' + \delta\delta' \end{bmatrix}. \tag{5.17}$$

Exercise: Given the matrices M_S and M_T as above, compute M_{TS} and give an example to show that $M_{TS} \neq M_{ST}$.

The algebraic properties on $\mathcal{L}(\mathbb{R}^2)$, given in relations A(i)-A(vii) in Section 5.3, can similarly be translated for the corresponding matrices. From this it should be clear that

$$M_0 = \begin{bmatrix} 0 & 0 \\ 0 & 0 \end{bmatrix} \text{ and } M_{id} = \begin{bmatrix} 1 & 0 \\ 0 & 1 \end{bmatrix}, \tag{5.18}$$

where id stands for the (linear) identity map on $\mathbb{R}^2$ (see Example (h) in Section 1.6). The distinguished matrix M_{id} is called the identity matrix and denoted by I.

The next natural question that arises is that if $S \in \mathcal{L}(\mathbb{R}^2)$, given by (5.16), is one-to-one (and therefore bijective by the last theorem in Section 5.4), then what is the matrix $M_{S^{-1}}$? By (5.15) and (5.18) we have that $M_S M_{S^{-1}} = M_{id} = I$, and if we set for convenience $S^{-1} = T$ so that S^{-1} is represented by the expression for T in (5.16), we conclude that $\alpha\alpha' + \beta\gamma' = \gamma\beta' + \delta\delta' = 1$ and $\alpha\beta' + \beta\delta' = \gamma\alpha' + \delta\gamma' = 0$. It remains to solve (by elimination) for the four real numbers $(\alpha', \beta', \gamma', \delta')$ from the above four equations to find that (i) this is possible if and only if $\Delta(S) \equiv \alpha\delta - \beta\gamma \neq 0$ (see also the proof of the last theorem in Section 5.4) and (ii) in such a case, the matrix

$$M_T = M_{S^{-1}} = \frac{1}{\Delta(S)} \begin{bmatrix} \delta & -\beta \\ -\gamma & \alpha \end{bmatrix}.$$

The expression $\Delta(S)$ is called the *determinant* of the matrix M_S. The verification of (i) and (ii) above is left as an **exercise**.

5.6 Linear Transformations and Geometry

So far we have studied only the algebraic properties of linear transformations and their associated matrices. Here we introduce some geometric concepts in $\mathbb{R}^2$ and study their consequences. For this it is useful to consider the elements of $\mathbb{R}^2$ as vectors in a 2-dimensional plane.

Inner Product and Length

Let $\boldsymbol{x} = (a, b)$ and $\boldsymbol{y} = (c, d)$ be two vectors in $\mathbb{R}^2$. We define the *inner product* of $\boldsymbol{x}$ and $\boldsymbol{y}$, denoted as $\langle \boldsymbol{x}, \boldsymbol{y} \rangle$, by

$$\langle \boldsymbol{x}, \boldsymbol{y} \rangle = ac + bd, \tag{5.19}$$

and the length $\|x\|$ of $x = (a, b)$ from the origin, called the norm of x, by

$$\|x\| = (a^2 + b^2)^{\frac{1}{2}}. \tag{5.20}$$

Obviously $\|x\| = \langle x, x\rangle^{\frac{1}{2}}$. It can also be shown that $\langle x, y\rangle = \|x\| \; \|y\| \cos\theta$, where θ is the angle between the vectors x and y (**Exercise**). (Hint: Use the expression for $\cos\theta$ from coordinate geometry). We say that the non-zero vectors x and y are *orthogonal* if

$$\langle x, y\rangle = 0.$$

Then we have the following theorem, the proof of which is left as an **exercise**.

THEOREM: *Let $x, y, z \in \mathbb{R}^2$ and let $\alpha \in \mathbb{R}$. Then*

(i) $\langle x, y\rangle = \langle y, x\rangle$.

(ii) $\langle \alpha x, y\rangle = \langle x, \alpha y\rangle$.

(iii) $\langle x, y + z\rangle = \langle x, y\rangle + \langle x, z\rangle$.

(iv) $\langle x, x\rangle = 0$ *if and only if* $x = \mathbf{0}$.

(v) *If x and y are orthogonal, then x and y are linearly independent.*

Exercises

Prove the following:

(i) $|\langle x, y\rangle| \leq \|x\| \; \|y\|$.

(ii) $\|x + y\|^2 = \|x\|^2 + 2\langle x, y\rangle + \|y\|^2$.

(iii) $\|x + y\| \leq \|x\| + \|y\|$.

Exercises

Prove the following:

(i) To each linear transformation T, there corresponds a positive number k such that $\|Tx\| \leq k\|x\|$ for all x in $\mathbb{R}^2$. (The smallest number k satisfying the above inequality is called the norm of T and is denoted by $\|T\|$.)

(ii) Show that the matrix M_T of a linear transformation T with respect to the basis $e_1 = (1, 0)$ and $e_2 = (0, 1)$ is

$$\begin{bmatrix} \langle Te_1, e_1\rangle & \langle Te_2, e_1\rangle \\ \langle Te_1, e_2\rangle & \langle Te_2, e_2\rangle \end{bmatrix}.$$

Subspace: Let $\mathcal{H}$ be a subset of $\mathbb{R}^2$. $\mathcal{H}$ is said to be a subspace of $\mathbb{R}^2$ if

(i) $x + y$ is in $\mathcal{H}$ whenever x and y are in $\mathcal{H}$ and

(ii) αx is in $\mathcal{H}$ whenever x is in $\mathcal{H}$ and α is any real number.

Remark: $\mathbf{0}$ is in every subspace of $\mathbb{R}^2$. Also, clearly $\mathbb{R}^2$ itself and the singleton set $\{\mathbf{0}\}$ are subspaces of $\mathbb{R}^2$. Similarly, the sets $\{(x, 0) | x \in \mathbb{R}\}$ and $\{(x, x) | x \in \mathbb{R}\}$ are subspaces of $\mathbb{R}^2$ (verify !). In particular, the last two subspaces correspond geometrically to the x-axis and the straight line through the origin with slope 1 respectively.

Exercise: Let T be a linear transformation and let $\mathcal{H}$ be a subspace of $\mathbb{R}^2$. Then the image under T of $\mathcal{H}$ is also a subspace of $\mathbb{R}^2$.

Eigenvectors and Eigenvalues

Let T be a linear transformation on $\mathbb{R}^2$. The real number α is called an eigenvalue of T if there exists a non-zero vector x in $\mathbb{R}^2$ such that

$$Tx = \alpha x.$$

In this event x is called an eigenvector of T corresponding to α. If M_T is the matrix of the transformation T, x and α are also referred to as eigenvector and eigenvalue of M_T respectively.

Exercises

(i) Let α_1, α_2 be distinct eigenvalues of T and let x_1, x_2 be the corresponding eigenvectors. Then prove that x_1 and x_2 are linearly independent. Show that there can be at most two distinct eigenvalues of T. Give an example of T which has only one eigenvalue.

(ii) Prove that the collection of all eigenvectors that correspond to an eigenvalue α of a linear transformation T form a subspace.

The notions of eigenvalues and eigenvectors are of fundamental importance in atomic physics, since the energy levels of atoms and molecules turn out to be given by the eigenvalues of certain linear transformations but in infinitely many dimensions (see Section 5.8). Also, the analysis of vibrations and the stability analysis of many mechanical and economic systems require finding the eigenvalues and eigenvectors of certain linear transformations.

Transpose of a Linear Transformation

If $T \in \mathcal{L}(\mathbb{R}^2)$, then we can associate with it another transformation T^* (called the *transpose* of T) on $\mathbb{R}^2$ by

$$\langle T\boldsymbol{x}, \boldsymbol{y}\rangle = \langle \boldsymbol{x}, T^*\boldsymbol{y}\rangle \quad \text{for every pair } \boldsymbol{x}, \boldsymbol{y} \in \mathbb{R}^2. \tag{5.21}$$

It can be easily shown that T^* is also a linear transformation and that if

$$M_T = \begin{bmatrix} \alpha & \beta \\ \gamma & \delta \end{bmatrix}, \tag{5.22}$$

then

$$M_{T^*} = \begin{bmatrix} \alpha & \gamma \\ \beta & \delta \end{bmatrix}. \tag{5.23}$$

Conversely, if we have a matrix as given by (5.22) and we treat it as a linear transformation, then the transpose of this linear transformation is the linear transformation given by the matrix in (5.23).

Exercise: If $T \in \mathcal{L}(\mathbb{R}^2)$ is given by M_T as in (5.22), find T^* and its associated matrix and thereby verify that $T^* \in \mathcal{L}(\mathbb{R}^2)$.

Symmetric Linear Transformations

A linear transformation T is said to be *symmetric* if $T = T^*$ or equivalently,

$$\langle Tx, y\rangle = \langle x, Ty\rangle \quad \text{for all } x \text{ and } y \text{ in } \mathbb{R}^2. \tag{5.24}$$

Examples

(a) $T(a, b) = (a, 0)$.

(b) $T(a, b) = (a+b, a+b)$.

Exercises

(i) Let T be given by its matrix in (5.22). Show that T is symmetric if and only if $\beta = \gamma$.

(ii) Let T be a symmetric transformation and let α, β be its two distinct eigenvalues with corresponding eigenvectors $\boldsymbol{x}, \boldsymbol{y}$. Then show that $\langle \boldsymbol{x}, \boldsymbol{y} \rangle = 0$ and hence $\boldsymbol{x}$ and $\boldsymbol{y}$ are linearly independent. Thus eigenvectors corresponding to two distinct eigenvalues of such a T form a basis. Are $\boldsymbol{x}, \boldsymbol{y}$ linearly independent even if T is not symmetric?

(iii) Let $\mathcal{H}$ be any non-empty set in $\mathbb{R}^2$ and let $\mathcal{H}^{\perp} = \{\boldsymbol{y}$ in $\mathbb{R}^2$ such that $\langle \boldsymbol{x}, \boldsymbol{y} \rangle = 0$ for all $\boldsymbol{x}$ in $\mathcal{H}\}$. Show that $\mathcal{H}^{\perp}$ is a subspace of $\mathbb{R}^2$.

(iv) If T is a symmetric transformation and $\mathcal{H}$ is a subspace of $\mathbb{R}^2$, show that T maps $\mathcal{H}$ into itself if and only if T maps $\mathcal{H}^{\perp}$ into $\mathcal{H}^{\perp}$.

(v) Let $\boldsymbol{x}(\neq \boldsymbol{0}) \in \mathbb{R}^2$ and let $T \in \mathcal{L}(\mathbb{R}^2)$. Suppose furthermore that T maps the subspace spanned by $\boldsymbol{x}$ into itself. Show that $\boldsymbol{x}$ is an eigenvector of T.

THEOREM: *Every symmetric transformation T has two eigenvalues (not necessarily distinct) and there exist two mutually orthogonal eigenvectors corresponding to these eigenvalues.*

PROOF: Let T be a symmetric element of $\mathcal{L}(\mathbb{R}^2)$. Then T has an eigenvalue λ if and only if there is a non-zero vector $\boldsymbol{x}$ such that $T\boldsymbol{x} = \lambda\boldsymbol{x}$, or equivalently, $(T - \lambda.id)\boldsymbol{x} = 0$. This happens if and only if $(T - \lambda.id)$ is not one-to-one, which is true if and only if $(M_T - \lambda I)$ is not invertible. But $(M_T - \lambda I)$ is not invertible if and only if $\Delta(M_T - \lambda I) = 0$ (see Section 5.5). Using the fact that T is symmetric, we have that $\Delta(M_T - \lambda I) = ((\alpha - \lambda)(\gamma - \lambda)) - \beta^2$. Thus λ is an eigenvalue of T if and only if it is a root of the quadratic equation $(\alpha - \lambda)(\gamma - \lambda) - \beta^2 = 0$. Hence the two eigenvalues of T are given by

$$\lambda_1 = \{\alpha + \gamma + ((\alpha - \gamma)^2 + 4\beta^2)^{\frac{1}{2}}\}/2,$$

$$\lambda_2 = \{\alpha + \gamma - ((\alpha - \gamma)^2 + 4\beta^2)^{\frac{1}{2}}\}/2.$$

Note that $\lambda_1 = \lambda_2$ if and only if $\alpha = \gamma$ and $\beta = 0$.

To complete the proof we have to show that there are two eigenvectors which are orthogonal to each other. If the eigenvalues are distinct, then this follows by exercise (ii) above. In the case when T has only one eigenvalue, we have seen that $\alpha = \gamma, \beta = 0$, which implies that $M_T = \alpha I$ or $T = \alpha.id$ and in such a case, every non-zero vector in $\mathbb{R}^2$ is an eigenvector. Therefore we can choose any two orthogonal vectors as eigenvectors. ∎

Exercise: (i) Let θ be a fixed number such that $0 < \theta < 2\pi$. Let T be the mapping from $\mathbb{R}^2$ to $\mathbb{R}^2$ that rotates the vector x by an angle θ in the counterclockwise direction about the origin. Show that T is a linear transformation and compute the matrix M_T. Prove that T preserves length, i.e., $\|Tx\| = \|x\|$ for all x in $\mathbb{R}^2$.

5.7 Applications

We shall consider a few simple applications of the theory that we have developed above.

Linear Equations

Let us suppose that we have a system of two linear equations in two real unknowns a, b:

$$\alpha a + \beta b = u,$$

$$\gamma a + \delta b = v.$$

In the language of linear transformations in $\mathbb{R}^2$, this can be rewritten as

$$Tx = z, \tag{5.25}$$

where $x = (a, b), z = (u, v)$ and T is given by (5.12).

Thus, solving the system of two linear equations given above is equivalent to solving for an element $x \in \mathbb{R}^2$ satisfying (5.25). This problem has already been addressed in the proof of the last theorem in Section 5.4.

Matrices and Electrical Networks

Consider the following circuit diagram (Figure 5.1).

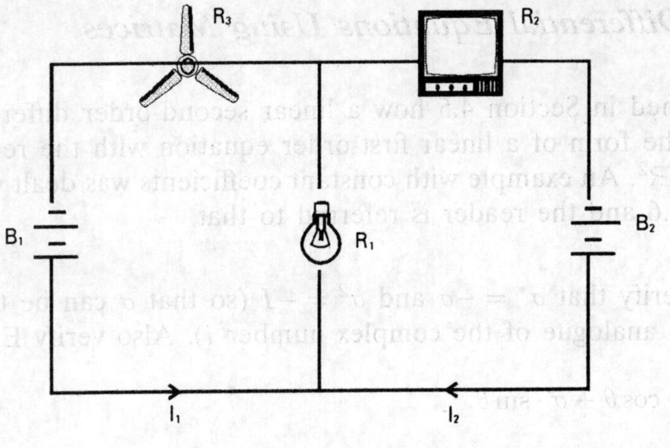

Figure 5.1

Let us assume that the voltage provided by the left hand battery source B_1 is V_1 and the voltage provided by the right hand battery source B_2 is V_2. Let the light bulb provide a resistance R_1, the television set provide a resistance R_2 and the fan provide a resistance R_3, all measured in ohms. Let us further suppose that the current across the terminals of the battery B_1 is I_1 and that the current across B_2 is I_2. By Kirchoff's law of electrical networks, we can determine the current in the various parts of the network and conclude that the current across the fan is I_1, the current across the television set is I_2 and the current across the bulb is $I_1 + I_2$. Now the voltage across the terminals of B_1 is the sum of the voltages across the fan and the bulb. By Ohm's law, and neglecting the internal resistance of the batteries,

$$V_1 = R_3 I_1 + R_1(I_1 + I_2).$$

In exactly the same way we can figure out that

$$V_2 = R_2 I_2 + R_1(I_1 + I_2).$$

We can now put the two equations down as the single matrix equation

$$\begin{bmatrix} R_1 + R_3 & R_1 \\ R_1 & R_1 + R_2 \end{bmatrix} \begin{bmatrix} I_1 \\ I_2 \end{bmatrix} = \begin{bmatrix} V_1 \\ V_2 \end{bmatrix}.$$

Now we can solve for I_1, I_2 given V_1 and V_2.

Solution of Differential Equations Using Matrices

We have explained in Section 4.5 how a linear second-order differential equation can be cast in the form of a linear first-order equation with the required solution taking values in $\mathbb{R}^2$. An example with constant coefficients was dealt with in Remark (b) of Section 4.6 and the reader is referred to that.

Exercise: Verify that $\sigma^* = -\sigma$ and $\sigma^2 = -I$ (so that σ can be thought of as a two-dimensional analogue of the complex number i). Also verify Euler's formula

$$\exp \sigma\theta = I \cdot \cos\theta + \sigma \cdot \sin\theta.$$

5.8 Epilogue

All the algebraic properties of $\mathbb{R}^2, \mathcal{L}(\mathbb{R}^2)$ and of associated matrices carry over naturally to n-dimensions. Since $\mathbb{R}$ has been defined, we define $\mathbb{R}^n$ inductively for $n > 1$, i.e., we can think of $\mathbb{R}^n$ as simply the collection of all ordered n-tuples of reals, $(\alpha_1, \cdots, \alpha_n)$. Clearly, we can see that addition and scalar multiplication can be defined on $\mathbb{R}^n$ in exactly the same way as in $\mathbb{R}^2$, and that all the properties listed in A(i)–A(iv) and M(i)–M(v) remain valid. Similarly, the definition of a linear transformation as well as all of its properties carry over without any change. The next natural question that arises is: does $\mathbb{C}^n$, the collection of all n-tuples $(z_1, \ldots, z_n)$ of complex numbers, have a structure similar to that of $\mathbb{R}^n$? The difference here is that the scalar multiplication has to be by complex numbers. With this modification, the properties A(i)–A(iv) and M(i)–M(v) remain true. Linear transformations can be defined exactly as before. The theory here is much richer than it is for $\mathbb{R}^n$, for example, every linear transformation will have an eigenvalue unlike the case in $\mathbb{R}^2$. One can also ask: why not extend these ideas to the class of all sequences $(\alpha_1, \alpha_2, \ldots, \alpha_n, \ldots)$ constituting what we may name as $\mathbb{R}^\infty$ or $\mathbb{C}^\infty$ as the case may be? Then, the addition and scalar multiplication can be defined as before. Another example we can consider is that of $\mathcal{F}(\mathbb{R})$, the class of all real-valued functions on $\mathbb{R}$. Let a be a fixed real number and let T be a transformation from $\mathcal{F}(\mathbb{R})$ to $\mathcal{F}(\mathbb{R})$ defined by $(Tf)(x) = f(x + a)$ for all real numbers x. We can first check that if addition and scalar multiplication are defined as

(i) $(f + g)(x) = f(x) + g(x)$,

(ii) $(\alpha f)(x) = \alpha f(x)$,

then properties similar to A(i)–A(iv) and M(i)–M(v) in Section 5.2 hold, and T is a linear transformation on $\mathcal{F}(\mathbb{R})$. Similarly, let $\mathcal{P}$ denote the class of all polynomials of degree less than or equal to $n (n > 1$ and fixed), with addition and scalar multiplication being defined as in (i) and (ii) above. Define the differentiation map D from $\mathcal{P}$ to $\mathcal{P}$ as

$$D(a_0 + a_1x + a_2x^2 + \ldots + a_nx^) = a_1 + 2a_2x + 3a_3x^2 + \ldots + na_nx^{n-1}.$$

We can easily check that D is a linear transformation on $\mathcal{P}$.

Exercise: Verify the properties A(i)–A(iv) and M(i)–M(v) for $\mathbb{R}^n, \mathcal{F}(\mathbb{R})$ and $\mathcal{P}$ and the linearity properties of the transformation T and D described above.

Suggested Reading

1. *Finite Dimensional Vector Spaces* by P.R.Halmos, Springer Verlag 1993.

2. *Linear Functions and Matrix Theory* by B.Jacob, Springer Verlag 1995.

Chapter 6

Probability and Statistics

6.1 Introduction

Most students understand the meaning of *probability of an event* much before they formally attend a course on probability and statistics. The *toss of a coin* as a means to decide which of the two team captains gets to choose whether to bat first or field first in a game of cricket is a common situation and it is accepted that this is a fair way to decide. Also as children, most students would have played games based on chance, such as snakes and ladders and ludo, which involve the throw of a six-faced die.

However, many students have difficulties when one tries to define the notions of *random events* and their associated probabilities. Some would think that these are applicable only to the toss of a coin, throw of a die or drawing a card from a deck of cards, and hence is relevant only to casinos and gamblers and has no place in real life applications. Some students could even argue that the outcome of the toss of a coin is not quite a random event: if we could write down all the parameters such as shape and weight of the coin, the amount of force applied and the points where it is applied while tossing the coin, the height above the ground level from where it is tossed, the characteristics of the wind at that time, etc., then the equations governing the motion of the coin can be written down and solved and the exact trajectory of the coin can be computed, from which we can tell if the coin will turn up heads or tails. Thus really speaking, the outcome of the toss of a coin is deterministic, though it is difficult to measure the quantities listed above and then obtain the exact trajectory.

Einstein is supposed to have said that God does not toss a die to decide as to which way an electron should spin – and yet students of chemistry are taught that an electron spins in clockwise direction with probability .5 and anticlockwise direction with probability .5. How does one justify or interpret this? It can be argued that though today there is no explanation for deciding as to which way (clockwise or anticlockwise) a given electron will spin in an atom, it is possible that some day in future, someone will come up with a complete explanation of the phenomenon. In any case we cannot rule out the possibility of such an explanation.

Thus, events that appear to be random or are considered to be random today may not be so tomorrow. How can we have a framework where this is not logically inconsistent?

One such framework is to think of the notion of randomness as a mathematical model for events whose outcome is not completely specified – leaving the possibility open that the model can change from time to time depending upon the understanding of the phenomenon, and also from person to person (rather, observer to observer) depending upon the information they have about the event. By the term mathematical model we mean a mathematical description relating various quantities of interest in a given physical context.

Immediately two questions arise. How can we model an event if its outcome is uncertain? And, why should we model such events?

Let us look at the first question. *Can we mathematically model uncertain events?* Once again, we have to go to real life and we can come up with many examples where the outcome of an event seems uncertain but on repeated trials, a pattern emerges. Thus if we toss a coin or a die repeatedly, we can observe a pattern. The oldest such example is child birth: at individual (micro) level, the event *the child is a girl* is uncertain, yet it has been observed that the proportion of girls amongst the number of children born in a city in a month is approximately $\frac{1}{2}$! This has been observed across continents over several centuries. Thus at micro level (*i.e.,* at the level of individual births), the event is unpredictable but at macro level (*i.e.,* at the level of all births in a city in a given month), it generates patterns. Perhaps, we can use these patterns to come up with a model.

Let us now look at the second and more important question – why should we model uncertainty at all (if we are not interested in gambling)? We can think of many situations where we are not certain about the outcome of an event, but we need to decide on a course of action. Suppose there are two routes to reach a destination, the first one is shorter, but goes through the city and is known to have traffic jams, while the second one goes along the outskirts and is usually without traffic jams. Mr. A has an appointment at 10:00 a.m. for a job interview and he is delayed at home due to reasons beyond his control. He knows that if he takes the safer route that takes longer, he will reach his destination only around 10:15 a.m., whereas he could just reach in time via the shorter route, if there are no traffic jams (if there is indeed a jam he could be delayed by up to an hour, say). Now, here common sense tells us that he should take the risk and take the shorter route – for, with the safer route, he is sure to miss his appointment. However, if he were going instead for a 3-hour examination, perhaps he should choose the safer but longer route and be 15 minutes late rather than risk the possibility of an hour's delay.

In such simple cases we can (and do) use common sense and come up with a solution. If we could mathematically model the uncertainty, then the principles behind common sense in simple cases can be isolated and applied to more complex situations.

We can come up with many situations where we do not quite have an exact understanding of the phenomenon, but we use experience – our own or someone else's to make decisions. For example, how much water to give to a given plant? Which bus routes to choose to go from a point A to a point B at a given time of the day? In the second example, we may have several choices and we would choose one, which we believe will be less crowded or will get us to the destination faster. We do not know for sure answers to these questions, but we may have information in the form of experience, our own or of others. Such information is called empirical evidence.

Even in crucial areas such as medicine, a lot of the knowledge that we have is in the form of empirical evidence. In many cases the choice of the medicine itself could be based on century-old practices in some part of the world. In some other cases, the medicine could be arrived at by an understanding of the body chemistry, but the assessment of the side effects is invariably based on empirical evidence—based on clinical trials where the medicine is administered to some animals or humans and they are closely monitored.

To sum up, there are many situations where we need to make decisions under uncertainty and if we could evolve mathematical models for the underlying phenomenon, it would be helpful in decision making.

At this point, it is recommended that the students conduct an experiment for themselves. Toss a coin 4 times (or take 4 coins and toss them together). Note the number of heads observed. Repeat this 256 times. Let n_0 be the number of times (out of 256) when no head was observed; n_1, n_2, n_3, n_4 be the number of times that 1, 2, 3 and 4 heads respectively were observed (so that $n_0 + n_1 + \cdots + n_4 = 256$). Compare your answers with those of a friend or friends who performed the same experiment independently of you. You will see that the values of $n_0, n_1, ..., n_4$ observed by different persons are close to each other. It will be seen that the values will be close to 16, 64, 96, 64, 16 respectively—if not for everybody, for most of the students in the class.

Now we come up with a mathematical model that will explain this.

6.2 Randomness as a mathematical model for uncertainty

We observed in the previous section that we can view probability as a mathematical model for events whose outcome is uncertain. Let us start with some thoughts on what we want this model to be. Consider an experiment whose outcome is uncertain. We might be interested particularly in the occurrence of one or more specified outcomes from the set of all possible outcomes. Every such set (a subset of the set of all outcomes) is called an event. For each event A associated with this experiment, we assign a number $P(A)$ between 0 and 1 and call it the probability of occurrence of A. Heuristically, we want the probability $P(A)$ of occurrence of A to reflect the relative frequency with which A occurs in repeated trials of the experiment. (Relative frequency of occurrence of A means the ratio of the number of times A occurs to the total number of trials.) This will be our guiding principle in constructing the model, i.e., in assigning probabilities.

We will begin by considering several experiments where this guiding principle and inherent symmetries in the experiment lead us to a model. Later, we use these models as building blocks in more complex situations.

Example 1 (Toss of a coin): Consider the familiar experiment of tossing of a fair coin. In view of the symmetries of the coin, we do not expect the relative frequency of *heads* to be more than the relative frequency of *tails* and vice versa. Thus, if the relative frequencies of *heads* and *tails* do stabilise, i.e., converge to a number as the number of trials increase, the two numbers must be the same and add to 1 (which means each of them is .5). So it seems reasonable to model the probability associated with the toss of a fair coin as:

Probability of a head turning up = .5.

Probability of a tail turning up = .5.

Example 2 (Throw of a die): Consider the throwing of a symmetric die. The die as usual has dots marked on each of its six sides, with one side having one dot, another side having two dots and so on. Let E_1 denote the event that the side with one dot turns up. Similarly, events $E_2, ..., E_6$ are defined. As in Example 1, we can argue that if the relative frequencies of $E_1, E_2, ..., E_6$ do stabilise, they should stabilise to $\frac{1}{6}$. So we may model the probability of each of the events $E_1, E_2, ..., E_6$ as $\frac{1}{6}$. Another way of putting this is: in view of the inherent symmetry, the six outcomes are equally likely.

Example 3 (Drawing a ball from an urn): Consider an urn containing 10 balls, numbered 1,2,..., 10 respectively, and that balls with numbers 1,2,3,4 are red in colour and the remaining 6 are green. Suppose that apart from the distinct numbers and differences in colour, the balls are identical in all other aspects such as size, shape, weight, etc. The experiment consists of mixing the balls in the urn and drawing a ball out of the urn (the ball is picked up without noting the colour). Denoting by $F_1, F_2, ..., F_{10}$ the events that the ball drawn carries the number 1,2,..., 10 respectively, we could argue as in the two previous experiments that it is reasonable to model these events as equally likely, so that we have $P(F_i) = .1,\ i = 1, 2, \ldots, 10$ as our model. If A denotes the event that the ball drawn is of red colour, then it is clear that the relative frequency of occurrence of A is the sum of relative frequencies of occurrences of F_1, F_2, F_3 and F_4, and thus we can model the probability of A as

$$P(A) = P(F_1) + P(F_2) + P(F_3) + P(F_4) = .4.$$

Now, let us consider the following variation. Suppose that the balls do not have any numbers. There are 4 red and 6 green balls in the urn. This time, we do not have symmetry: the events of drawing a red ball and of drawing a green ball are not equally likely. We can pretend that the balls carry numbers, the 4 red balls having numbers 1,2,3,4 and the green balls having numbers 5,6,7,8,9,10 and then we are in the situation discussed earlier. So even when the balls do not have numbers, it would be reasonable to model the probability of drawing a red ball as .4 and that of drawing a green ball as .6.

Example 4 (Drawing a ball from an urn, II): Suppose an urn has N red balls and M green balls, identical in all other aspects. The experiment consists of mixing the balls and drawing a ball from the urn (without looking at the colour). Following the discussion in the previous example, it is reasonable to model the probabilities as:

$$\text{P(ball drawn is red)} = \frac{N}{N+M}.$$

$$\text{P(ball drawn is green)} = \frac{M}{N+M}.$$

Example 5 (Repeated trials of drawing a ball from an urn): Suppose we perform the experiment in Example 4 repeatedly: at each draw, we note the colour of the ball after it is drawn and put it back in the urn, mix the balls and draw the next ball and so on. This way we draw n balls. We are interested in the event that exactly k out of n balls are of red colour and $n-k$ are of green colour.

To come up with a model for probabilities of these events, let us assume that the balls have an invisible tag on them: the N red balls have tags $1, 2, ..., N$ and the green balls have tags $N+1, N+2, ..., N+M$. At any given trial, the $(N+M)$ events: the tag on the chosen ball is $i (1 \le i \le N+M)$ can be taken to be equally likely. Further, the outcomes of the previous draws have no bearing on the subsequent draws as the balls are put back in the urn. Thus, all the events $E(i_1, i_2, ..., i_n)$ that the successive balls drawn have tags $i_1, i_2, ..., i_n$ can be taken to be equally likely. There are $(N+M)^n$ possibilities and hence the probability of each of these events can be taken to be $\frac{1}{(N+M)^n}$. It remains to count (as in Example 3) as to how many of these $(N+M)^n$ events correspond to the event of interest; namely, k red and $(n-k)$ green balls are drawn.

For every choice of k distinct integers $a_1, a_2, ..., a_k$ out of $\{1, 2, ..., n\}$, we will have $N^k M^{n-k}$ events $E(i_1, i_2, ..., i_n)$ such that $i_j \leq N$ if $j = a_1, a_2, ..., a_k$ and $i_j > N$ otherwise. These correspond to drawing red balls exactly at the $a_1^{th}, a_2^{th}, ..., a_k^{th}$ draw and drawing green at other draws. Since we can choose $a_1, \ldots, a_k$ in $\binom{n}{k}$ ways, it follows that the total number of outcomes that correspond to drawing k red and $n - k$ green balls is

$$\binom{n}{k} N^k M^{n-k},$$

and hence it is reasonable to model

$$\begin{aligned} P(\text{drawing exactly } k \text{ red balls in } n \text{ trials}) &= \binom{n}{k} \frac{N^k M^{n-k}}{(N+M)^n} \\ &= \binom{n}{k} p^k (1-p)^{n-k}, \end{aligned}$$

where $p = N/(N + M)$ is the probability of drawing a red ball in a single trial.

Example 6 (Repeated tosses of a coin): Suppose a coin (assumed to be symmetrical) is tossed n times and we are interested in the event: exactly k *heads* and $n - k$ *tails*. We can draw an analogy with the previous example: taking $N = 1 = M$, drawing a red ball is identified with head and that of a green ball is identified with tail. Thus, we can take the model as

$$P(\text{exactly } k \textit{ heads} \text{ in } n \text{ tosses}) = \tbinom{n}{k} \cdot 2^{-n}.$$

Let us stop and reflect for a moment on what we have observed so far. The main idea in the preceding examples was to invoke inherent symmetry to model certain outcomes as equally likely, and thus having equal probabilities. Further, if an event is made up of a certain number of these outcomes, then the probability of the event is the sum of probabilities of the outcomes constituting it.

Thus, if an experiment results in N outcomes that can be taken as equally likely and if an event A consists of n_1 of these N outcomes, then we can take

$$P(A) = \frac{n_1}{N}.$$

If B is another event consisting of n_2 of these outcomes such that A and B have no outcome in common, then $P(B) = \frac{n_2}{N}, P(A \text{ or } B) = \frac{(n_1+n_2)}{N}$ and hence we have

$$P(A \text{ or } B) = P(A) + P(B),$$

if A, B are mutually disjoint, i.e., A and B cannot occur simultaneously.

If C, D are events consisting of m_1, m_2 outcomes respectively and out of these, k outcomes are common, then the event C or D consists of $m_1 + m_2 - k$ outcomes. Thus we get

$$\begin{aligned} P(C \text{ or } D) &= \frac{m_1 + m_2 - k}{N} \\ &= P(C) + P(D) - P(C \text{ and } D). \end{aligned}$$

Law of large numbers

Let us now examine what happens when we repeat an experiment a large number of times. Let us go back to Example 5 (where we are going to take n to be very large). Let X_n denote the number of red balls drawn in the n trials. We have seen that

$$P(X_n = k) = \binom{n}{k} p^k (1-p)^{n-k},$$

where $p = \frac{N}{N+M}$ is the probability of drawing a red ball at the first draw. We expect that as n becomes large, the relative frequency $\frac{X_n}{n}$ of red balls in n trials is approximately equal to p. What is in fact true is that as n becomes large,

$$P\left(\left|\frac{X_n}{n} - p\right| \geq \varepsilon\right)$$

approaches zero for any given $\varepsilon > 0$. We give below a proof of this result (this is a version of the *law of large numbers* that we will dwell upon later).

Note that

$$\begin{aligned} \sum_{k=0}^{n} k \binom{n}{k} p^k (1-p)^{n-k} &= \sum_{k=1}^{n} \frac{n!}{(k-1)!(n-k)!} p^k (1-p)^{n-k} \\ &= \sum_{j=0}^{n-1} np \frac{(n-1)!}{j!(n-1-j)!} p^j (1-p)^{(n-1)-j} \\ &= np(p+1-p)^{n-1} = np. \end{aligned} \quad (6.1)$$

Similarly, we can prove that

$$\sum_{k=0}^{n} k(k-1)\binom{n}{k}p^k(1-p)^{n-k} = n(n-1)p^2$$

and hence that

$$\begin{aligned}\textstyle\sum_{k=0}^{n} k^2\binom{n}{k}p^k(1-p)^{n-k} &= n(n-1)p^2 + np \\ &= n^2p^2 + np - np^2. \qquad (6.2)\end{aligned}$$

Using (6.1) and (6.2), it follows that

$$\begin{aligned}\sum_{k=0}^{n}\left(\frac{k}{n}-p\right)^2\binom{n}{k}p^k(1-p)^{n-k} &= \sum_{k=0}^{n}\left(\frac{k^2}{n^2}-\frac{2kp}{n}+p^2\right)\binom{n}{k}p^k(1-p)^{n-k} \\ &= \frac{n^2p^2+np-np^2}{n^2} - \frac{2.np.p}{n} + p^2 \\ &= \frac{p(1-p)}{n}.\end{aligned}$$

Now

$$\begin{aligned}P(|\frac{X_n}{n}-p| \geq \varepsilon) &= \sum_{k:|\frac{k}{n}-p|\geq\varepsilon}\binom{n}{k}p^k(1-p)^{n-k} \\ &\leq \frac{1}{\varepsilon^2}\sum_{k:|\frac{k}{n}-p|\geq\varepsilon}(\frac{k}{n}-p)^2\binom{n}{k}p^k(1-p)^{n-k} \\ &\leq \frac{1}{\varepsilon^2}\frac{p(1-p)}{n} \leq \frac{1}{4\varepsilon^2 n}.\end{aligned}$$

As n becomes large, $\frac{1}{4\varepsilon^2 n}$ converges to zero and hence $P(|\frac{X_n}{n}-p| \geq \varepsilon)$ converges to zero as n becomes large.

Let us note that the proof given above rests only on our model that all the outcomes $E(i_1, i_2, ..., i_n)$ (that the ball with (invisible) tag i_1 is drawn at the first draw, tag i_2 at the second draw,. . ., tag i_n at the nth draw) are equally likely.

The result given above is a special case of the law of large numbers which we will state later. The law of large numbers is also commonly known as the *law of averages*.

6.3 Independence

The notion of independence plays a very special role in probability theory. It concerns the following problem: Suppose we have constructed probability models for two experiments (i.e., we have assigned probabilities for outcomes of each of the experiments). How do we go about assigning probabilities to events that depend upon both the experiments?

The simplest situation is when the two experiments have nothing to do with each other. For example, simultaneous toss of a coin and a die. In this case, using symmetry one could argue that it is reasonable to model all the $2 \times 6 = 12$ outcomes as equally likely. Another example is drawing a ball from an urn a few times, replacing the ball drawn and mixing the balls before drawing the next one.

Consider two experiments, with the first experiment yielding n_1 equally likely outcomes and second experiment resulting in n_2 equally likely outcomes. Let an event A_1 consist of m_1 out of the n_1 outcomes of the first experiment and A_2 consist of m_2 out of the n_2 outcomes of the second. Suppose that the outcome of the first experiment has no influence on the outcome of the second as in the two examples discussed above.

Now as seen above, we could model all the $n_1 \times n_2$ outcomes of the joint experiments as equally likely, and out of these $m_1 \times m_2$ outcomes make up the event (A_1 and A_2). Thus,

$$P(A_1 \text{ and } A_2) = \frac{m_1 \times m_2}{n_1 \times n_2} = P(A_1) \times P(A_2).$$

We can easily generalise this to k experiments: the ith experiment resulting in n_i equally likely outcomes out of which m_i constitute event A_i, $i = 1, 2, \ldots, k$. If the experiments are independent of each other, then the $n_1 \times n_2 \times \cdots \times n_k$ outcomes can be taken to be equally likely and $m_1 \times m_2 \times \cdots \times m_k$ of these constitute the event A_1 and A_2 and ... and A_k. We thus get

$$\begin{aligned} P(A_1 \text{ and } A_2 \text{ and } \cdots \text{and } A_k) &= \frac{m_1 \times \cdots \times m_k}{n_1 \times \cdots \times n_k} \\ &= P(A_1) \times P(A_2) \times \cdots \times P(A_k). \end{aligned}$$

Taking a cue from this, we adopt the following principle: if events $B_1, B_2, \ldots, B_k$ correspond to experiments that are independent of each other, and if we have assigned probabilities to $B_1, B_2, \ldots, B_k$, then it is reasonable to assign $P(B_1) \times P(B_2) \times \cdots \times P(B_k)$ as the probability to the event B_1 and B_2 and ... and B_k:

$$P(B_1 \text{ and } B_2 \text{ and ... and } B_k) = P(B_1) \times P(B_2) \times \cdots \times P(B_k).$$

Some concrete probability models

So far, we have been considering examples which can be associated with gambling. We chose to discuss them for the simple reason that they can be performed repeatedly, by anyone without any cost. Readers should try one or two of these experiments a large number of times and compare their observed relative frequencies and the probabilities assigned by our models.

Let us now turn to some concrete examples. Right from the time Madame Curie discovered radioactivity, numerous experiments have confirmed that the amount of radiation emitted by radioactive substances such as radium and uranium decays at an exponential rate. Also, the rate of decay does not depend upon the shape, size or temperature of the material. The amount of time it takes for the radioactive radiation to come down to half the level (known as half life) depends only on the substance. So at the macro level, (i.e., for the collection of molecules) one has a very good model for radioactivity, and it fits observed phenomenon—namely, if one has an amount a_0 of radium, say at time 0, then the amount a_t of radium at time t is given by

$$a_t = a_0 \exp(-\rho t),$$

where ρ is the rate of decay and is a constant. If τ is the half life of radium, then clearly, ρ and τ are related by

$$\exp(-\rho\tau) = \frac{1}{2}$$

so that $\rho\tau = \log_e 2$.

Can we construct a *micro* model for the behaviour of individual molecules that is consistent with the observed macro behaviour?

Since the rate of decay ρ (or the half life τ) does not depend on the shape or size of the substance, it would be reasonable to model that the individual molecules act independently of each other—for if there were interaction, the shape and size of the substance would affect the decay.

Let us try the following model: a molecule disintegrates in α units of time with probability p_α (and does not disintegrate with probability $1 - p_\alpha$).

Also, we stipulate that the event, *a given particle disintegrates*, is independent of what happens to the other particles.

Then we have

$$\begin{aligned} & P\,(\text{a given particle does not disintegrate till time } n\alpha) \\ &\quad = \prod_{j=0}^{n-1} P\,(\text{the particle does not disintegrate during } (j\alpha, j\alpha+\alpha)) \\ &\quad = (1 - p_\alpha)^n. \end{aligned}$$

If we start with say 1 gram of radium, it of course has a very large number of molecules, and then by our guiding principle the observed proportion of molecules that survive time $n\alpha$ should equal this probability. Thus we have

$$(1 - p_\alpha)^n = \exp(-\rho n\alpha)$$

or

$$1 - p_\alpha = \exp(-\rho\alpha).$$

Recalling that $\exp(-\rho\tau) = \frac{1}{2}$, we have

$$\begin{aligned} 1 - p_\alpha &= \exp(-\rho\alpha) \\ &= \left(\frac{1}{2}\right)^{\frac{\alpha}{\tau}} \end{aligned}$$

We thus get $p_\tau = \frac{1}{2}$. So, according to our model, a molecule survives its half life with probability $\frac{1}{2}$. Now the law of large numbers proved earlier gives that half the molecules present will survive the half life of the substance and thus our model is consistent with the observed macro phenomenon. Here, the presence of a very large number of molecules in any significant amount of the substance and the law of large numbers explain the observed deterministic rate of decay at the micro level.

Here we came across a distribution of probabilities that occurs in other contexts as well. If X denotes the number of unit time intervals that a molecule survives, writing $p = p_\alpha, q = 1 - p$, we get

$$P(X > n) = q^n$$

and thus

$$\begin{aligned} P(X = n) &= P(X > n - 1) - P(X > n) \\ &= q^{n-1} - q^n = q^{n-1}(1 - q) = q^{n-1}p. \end{aligned}$$

This is known as geometric distribution.

Let us consider another situation. A textile mill is producing cotton cloth—and inspite of its best efforts, it is observed that some defects do appear in the final product. We would like to construct a model for the number of defects per meter length of the cloth (the cloth has fixed width).

Let us divide a given one-meter length of cloth in N equal parts, where N is very large. Assuming that all efforts are made to remove any systematic patterns in the occurence of defects, it is reasonable to model that the probability of occurrence of defects in any given part is the same for the N parts and that the associated N events are independent. Since N is large, each part is small and we can ignore the possibility of occurence of two or more defects in one part. So if p_N is the probability of occurence of one defect in one of the N parts, then it follows that the total number of defects X in one-meter length has the following assignment of probabilities (under our model):

$$P(X = k) = \binom{N}{k} p_N^k (1 - p_N)^{N-k}.$$

This is called the binomial distribution.

Now the problem is to choose N and then to get p_N. First, note that

$$\begin{aligned}\sum_{k=0}^{N} kP(X = k) &= \sum_{k=0}^{N} k.\binom{N}{k} p_N^k (1 - p_N)^{N-k} \\ &= Np_N.\end{aligned}$$

The left hand side, namely, $\sum_{k=0}^{N} kP(X = k)$ can be interpreted as the expected value of X. We will later see that this expected value, λ, say, can be taken to be the average number of defects per meter length, where the average is taken over a large length. We thus get

$$p_N = \frac{\lambda}{N}.$$

Hence,

$$P(X = k) = \binom{N}{k} \left(\frac{\lambda}{N}\right)^k \left(1 - \frac{\lambda}{N}\right)^{N-k}$$

We can take the limit as N tends to infinity. Now

$$\begin{aligned}&\binom{N}{k} \left(\frac{\lambda}{N}\right)^k \left(1 - \frac{\lambda}{N}\right)^{N-k} \\ &= \frac{N(N-1)(N-2)\cdots(N-k+1)}{k!} \left(\frac{\lambda}{N}\right)^k \left(1 - \frac{\lambda}{N}\right)^{N-k}\end{aligned}$$

$$= \frac{\lambda^k}{k!}\left(1-\frac{\lambda}{N}\right)^N\left(1-\frac{1}{N}\right)\cdots\left(1-\frac{k-1}{N}\right)\left(1-\frac{\lambda}{N}\right)^{-k}$$

$$\to \frac{\lambda^k}{k!}e^{-\lambda}$$

as $N \to \infty$. We thus arrive at

$$P(X = k) = e^{-\lambda}\frac{\lambda^k}{k!}.$$

So in a given situation, the only thing we need to choose (or guess, or estimate) is λ.

This distribution of probabilities also occurs in many other situations, like the number of typographical errors per page in a book, or the number of accidents in a city, or the total number of telephone calls in a city in a given time interval. This distribution is called the Poisson distribution.

6.4 Random Variables and their expectations

In these examples we have come across variables such as the number of particles that disintegrate in a given length of time or the number of defects in a piece of cloth, whose outcome is uncertain, but we can model the probabilities of events associated with the variables. Such variables are called random variables and the probabilities of the events associated with a random variable is called its distribution.

To begin with, we will consider random variables taking values in the set $I \subseteq \mathbb{Z}$. Similarly, one can also consider random variables taking values in any countable set $\{a_n : n \geq 1\}$.

If we consider two random variables X, Y, then we must model

$$P(X = i, Y = j), \quad i, j \in I.$$

This is called the joint distribution of X, Y. The random variables X, Y are said to be independent if the events $\{X = i\}$, $\{Y = j\}$ are independent for each $i, j \in I$, and then one has

$$P(X = i, Y = j) = P(X = i)P(Y = j).$$

Similarly, random variables $X_1, X_2, \cdots, X_k$ are said to be independent if the events $\{X_1 = i_1\}$, $\{X_2 = i_2\}, \ldots, \{X_k = i_k\}$ are independent for $i_1, \ldots, i_k \in I$, and then

$$P(X_1 = i_1, X_2 = i_2, \ldots, X_k = i_k) = P(X_1 = i_1)P(X_2 = i_2)\cdots P(X_k = i_k).$$

For a random variable X (taking values in the set of integers I), its expected value $E(X)$ is defined by

$$E(X) = \sum_{i\in I} iP(X = i),$$

provided

$$\sum_{i\in I} |i|P(X = i) < \infty.$$

When I is a finite set, $E(X)$ always exists since the sum appearing above is a finite sum.

If $E(X)$ is defined, then the variance, $\mathrm{Var}(X)$ is defined as

$$\mathrm{Var}(X) = \sum_{i\in I}(i - E(X))^2 P(X = i).$$

If X and Y are a pair of random variables taking values in I with finite variances, then the *covariance* of X and Y is defined as

$$\mathrm{Cov}(X,Y) = E\{(X - E(X))(Y - E(Y))\}.$$

We will now state the *law of large numbers* for integer-valued random variables.

THEOREM: *Let $X_1, X_2, \ldots, X_n, \ldots$ be a sequence of independent random variables with*

$$P(X_k = i) = P(X_1 = i),\ k \geq 1,\ i \in I$$

and $E(X_1) = \mu$. Then

$$P\left(\left|\frac{X_1 + \cdots + X_n}{n} - \mu\right| > \varepsilon\right) \to 0$$

as $n \to \infty$.

A special case of this was proved earlier.

A sequence $X_1, X_2, \ldots, X_n, \ldots$ as in the law of large numbers stated above, is called an i.i.d. (independent and identically distributed) sequence. A typical case where an i.i.d. sequence appears is that of repeated observations on a given phenomenon.

Let us consider again the example of the number of defects X per meter length of cloth produced in a given textile mill. We have seen that a good model for X is

$$P(X = k) = e^{-\lambda}\frac{\lambda^k}{k!}, \quad k \geq 0.$$

Here, $E(X) = \sum_{k=0}^{\infty} kP(X = k) = \sum_{k=0}^{\infty} k.e^{-\lambda}\frac{\lambda^k}{k!} = \lambda$.

Let $X_1, X_2, \ldots, X_n$ denote the observed number of defects in one-meter length of cloth, chosen every hour for a few days. Then if n is large, $\frac{X_1+\cdots+X_n}{n}$ is close to λ with large probability. Thus, we can take $\hat{\lambda}_n = \frac{X_1+\cdots+X_n}{n}$ as an *estimate* of λ.

Let us now consider an opinion poll conducted in a city, where in an upcoming election there are two candidates, "A" and "B".

Suppose that an agency chooses people in the city, one by one, without following any pattern, and asks them if they would vote for "A" or "B". In this way, they speak to 400 voters and find that 244 out of these 400 would vote for "B" and 156 would vote for "A". From this can we conclude that "B" will win the election?

Firstly, any conclusion reached via such a sampling technique cannot be 100% correct; there would always be a possibility that the conclusion so reached is proven to be incorrect subsequently.

Suppose that the number of voters in the city is N and out of this, N_A would support "A" and $N_B = N - N_A$ would support "B". Let $X_1, X_2, \ldots, X_n$ be random variables denoting the response of the voters chosen, defined as $X_i = 1$ if the ith chosen person is a supporter of "B" and $X_i = 0$, otherwise. Assuming that the n persons are chosen at random, it is reasonable to model $X_1, \ldots, X_n$ as i.i.d. with $P(X_i = 1) = N_B/N$. Then as seen earlier, $Y_n = X_1 + \cdots + X_n$ has binomial distribution, with parameters $n, p = \frac{N_B}{N}$:

$$P(Y_n = k) = \binom{n}{k} p^k (1-p)^{n-k}.$$

Writing $p_n = \frac{Y_n}{n}$, which denotes the proportion of voters supporting "B" in the sample, we have

$$\begin{aligned} P(|p_n - p| \leq t) &= P(p - t \leq p_n \leq p + t) \\ &= P(n(p-t) \leq Y_n \leq n(p+t)) \\ &= \sum_{k=i_1}^{i_2} \binom{n}{k} p^k (1-p)^{n-k}, \end{aligned}$$

where i_1, i_2 are respectively the smallest and largest integers such that $n(p-t) \leq i_1 \leq i_2 \leq n(p+t)$.

Now if we choose $t = 0.1$ and n large enough such that

$$P(|p_n - p| \leq 0.1) \geq .95,$$

then we can say with 95% confidence that if we choose a random sample of size n, then the observed probability p_n and true proportion p would differ by at most 0.1. However, for a large n, exact computation of binomial probabilities and their sums involves a large amount of computation.

It was observed two centuries ago that for large n,

$$P\left(\left|\frac{\sqrt{n}(p_n - p)}{\sqrt{p(1-p)}}\right| \leq x\right)$$

is approximately equal to

$$\frac{1}{\sqrt{2\pi}} \int_{-\infty}^{x} e^{-y^2/2} dy$$

in the sense that the difference between the two terms goes to zero. Unfortunately,

$$\Phi(x) = \frac{1}{\sqrt{2\pi}} \int_{-\infty}^{x} e^{-y^2/2} dy$$

cannot be written explicitly, in closed form. However, tables of $\Phi(x)$ are available in various books and are also available with many statistical softwares. From this, it can be seen that

$$\Phi(1.96) - \Phi(-1.96) = .95.$$

Thus, we see that for large n,

$$P\left(\left|\frac{\sqrt{n}(p_n - p)}{\sqrt{p(1-p)}}\right| \leq 1.96\right) \simeq .95.$$

For $0 \leq p \leq 1$, $p(1-p) \leq 1/4$ and hence $\sqrt{4n}(p_n - p) \geq \frac{\sqrt{n}(p_n - p)}{\sqrt{p(1-p)}}$. Thus

$$\begin{aligned} P(|\sqrt{4n}(p_n - p)| \leq 1.96) &\geq P\left(\left|\frac{\sqrt{n}(p_n - p)}{\sqrt{p(1-p)}}\right| \leq 1.96\right) \\ &\simeq .95.. \end{aligned}$$

So we have, for large n,

$$P\left(|p_n - p| \leq \frac{1.96}{\sqrt{4n}}\right) \geq 0.95.$$

Thus for $n = 100$, we get $\frac{1.96}{\sqrt{4n}} = 0.098$ and thus

$$P(|p_n - p| \leq 0.098) \geq .95.$$

So if we draw a sample of size 100, we can say with 95% *confidence* that p_n and p would differ by at most 0.1.

Coming back to our example, we had $n = 400$ and then $\frac{1.96}{\sqrt{4n}} = 0.049$. Thus we can say with 95% confidence that p_n and p would differ by at most 0.049. We have observed $p_n = 0.61$ and thus with 95% confidence, we can say that p lies between $.61 \pm .049$ and thus "B" is likely to win.

In fact, $\Phi(2.54) - \Phi(-2.54) = 0.99$ and $\frac{2.54}{\sqrt{4n}} = .0635$, and hence we can say that

$$P(|p_n - p| \leq 0.0635) \geq .99 \quad (n = 400)$$

and so with 99% confidence, we can say that "B" will win.

The interval $(p_n - 0.0635, p_n + 0.0635)$ is called a 99% *confidence interval* for p and we have that with probability .99, this interval will contain the true p.

Another thing to note is that the population of the city does not come into the picture. To achieve a desired level of accuracy with a given degree of confidence, the sample size required does not depend upon the population to begin with.

The precise result on approximation of binomial probabilities mentioned above is

$$P\left(\frac{\sqrt{n}(p_n - p)}{\sqrt{p(1-p)}} \leq x\right) \to \Phi(x)$$

as $n \to \infty$. This is known as DeMoivre-Laplace central limit theorem.

We can interpret this as follows: for large n

$$\frac{\sqrt{n}(p_n - p)}{\sqrt{p(1-p)}}$$

behaves approximately as a random variable Z, where

$$P(Z \leq x) = \Phi(x).$$

Here, Z can take any value between $-\infty$ and ∞. Note that for all $\varepsilon > 0$

$$\begin{aligned} P(Z = x_0) &\leq P(Z \leq x_0) - P(Z \leq x_0 - \varepsilon) \\ &= \Phi(x_0) - \Phi(x_0 - \varepsilon) \end{aligned}$$

and hence using continuity of Φ we conclude that

$$P(Z = x_0) = 0, \text{ for all } x_0.$$

Such a random variable is called a *continuous* random variable. The probabilities for such a random variable Y are specified via a function $F(x)$, known as the distribution function of Y and is related to Y via

$$P(Y \leq x) = F(x).$$

If $F(x)$ can be written as

$$F(x) = \int_{-\infty}^{x} f(y)dy,$$

then f is called the density of Y.

Thus, DeMoivre-Laplace central limit theorem can be interpreted as: for large n, $\frac{\sqrt{n}(p_n - p)}{\sqrt{p(1-p)}}$ can be aproximately taken to be Z, where Z admits a density

$$f(x) = \frac{1}{\sqrt{2\pi}} e^{-x^2/2}, \quad -\infty < x < \infty.$$

The f appearing above is known as the density of a standard normal variate.

We have come across another example of a continuous random variable, though we were then dealing only with discrete random variables. We had seen that for a radioactive substance with half life τ,

$$P(\text{a given particle does not disintegrate till time } n\alpha) = \exp(-\rho n\alpha),$$

where $\rho = \frac{1}{\tau} \log_e 2$. Thus if X denotes the life span of the particle, then one has

$$P(X > n\alpha) = \exp\{-\rho n\alpha\}$$

or

$$P(X > t) = \exp\{-\rho t\}.$$

From this we get

$$\begin{aligned} P(X \leq t) &= 1 - P(X > t) \\ &= 1 - \exp\{-\rho t\} \\ &= \int_0^t \rho \exp\{-\rho s\} ds. \end{aligned}$$

Thus $f(s) = \rho \exp\{-\rho s\}, s \geq 0$ is the density of X.

In general, any function $f : \mathbb{R} \to \mathbb{R}$ such that $f(x) \geq 0$, $\int_{-\infty}^{\infty} f(x)dx = 1$ can be a density function of a continuous random variable. If

$$P(X \leq x) = \int_{-\infty}^{x} f(x)dx,$$

and $g : \mathbb{R} \to \mathbb{R}$ is a function such that

$$\int_{-\infty}^{\infty} |g(x)| f(x)dx < \infty,$$

we define

$$E[g(X)] = \int_{-\infty}^{\infty} g(x)f(x)dx.$$

As in the discrete case, the mean μ and variance σ^2 of a continuous random variable Z with density f are defined as

$$\mu = E[X] = \int_{-\infty}^{\infty} xf(x)dx.$$

If $\int_{-\infty}^{\infty} |x| f(x)dx < \infty$ and if μ is defined, then

$$\sigma^2 = E[(X-\mu)^2] = \int_{-\infty}^{\infty} (x-\mu)^2 f(x)dx.$$

Similarly, for a pair of random variables X and Y with finite variances, the covariance of X and Y is given by

$$\text{Cov}(X, Y) = E\{(X - E(X))(Y - E(Y))\}.$$

A sequence $Z_1, Z_2, \ldots, Z_n, \ldots$ is said to be a sequence of i.i.d. random variables if

$$P(Z_1 \leq z_1, Z_2 \leq z_2, \ldots, Z_k \leq z_k) = P(Z_1 \leq z_1)P(Z_2 \leq z_2) \cdots P(Z_k \leq z_k)$$

and

$$P(Z_i \leq z) = P(Z_1 \leq z).$$

DeMoivre-Laplace central limit theorem is true for a general sequence of i.i.d. random variables. The precise result is:

THEOREM: *Let $Z_1, Z_2, \ldots, Z_k, \ldots$ be a sequence of i.i.d. random variables. Suppose $\mu = E[Z_1]$ is defined and $\sigma^2 = E[(Z_1 - \mu)^2] < \infty$. Let $\bar{Z}_n = \frac{Z_1+Z_2+\ldots+Z_n}{n}$. Then*

$$P\left(\sqrt{n}\frac{\bar{Z}_n - \mu}{\sigma} \leq z\right) \rightarrow \int_{-\infty}^{z} \frac{1}{\sqrt{2\pi}} e^{-x^2/2} dx.$$

6.5 Elements of Statistical Inference

So far, we have seen how probability theory plays an important role in modelling physical phenomena under uncertainty. Probability theory also plays a crucial role in statistical inference, which is the core of statistical science. To motivate the discussion, we consider below two examples.

Example 7: Suppose it is desired to estimate the number of *Rehu* fish in a large pond, known to have a large number of *Rehu*. This is a typical problem of *statistical estimation*. How do we go about this problem? One of the ways of tackling this problem is as follows: A certain number of fish are caught in the first instance and the *Rehu* fish among these are marked with a red dye. Let the number of *Rehu* fish so marked be n_1. After marking the *Rehu*, these are released into the pond. Next day, a new catch of fish from the pond is made and it is observed that r of these are *Rehu*. Suppose among these r, k are with a red spot. We assume that the number of *Rehu* fish in the pond does not change between the first and second catches.

Suppose $n_1 = 88, r = 93, k = 19$. Then, we can say for sure that there are at least 88 + (93-19) = 162 *Rehu* fish in the pond. The total member of *Rehu* fish in the pond will be exactly 162 only if on the second day, all the unmarked *Rehu* fish are caught. Intuitively, this scenario is very unlikely and hence, we can conclude that most probably the number of *Rehu* fish in the pond is larger than 162.

If n is the (unknown) number of *Rehu* fish in the pond, then the probability that the second catch contains exactly k red-spotted *Rehu* fish is given by

$$p_k(n) = \frac{\binom{n_1}{k}\binom{n-n_1}{r-k}}{\binom{n}{r}}.$$

For the data at our disposal, we have $n_1 = 88, r = 93, k = 19$, so that

$$p_{19}(162) = 1.43 \times 2^{-28},$$

which is very small, as was suspected earlier.

A reasonable way to estimate the unknown 'parameter' n, given the quantities n_1, r, k would be to seek the particular value of n for which $p_k(n)$ attains its maximum, since for that n, our observation would have the largest probability. This method of estimation is called the method of *maximum likelihood*. For any given set of observations n_1, r, k, the value of n for which $p_k(n)$ is largest is called the maximum likelihood estimate of n, denoted by $\hat{n}$ (say).

To find $\hat{n}$, consider the ratio

$$R_n = \frac{p_k(n)}{p_k(n-1)} = \frac{(n-n_1)(n-r)}{(n-n_1-r+k)n}.$$

It is not hard to verify that

$$\begin{aligned} R_n \quad &> 1 \text{ if } \quad nk < n_1 r \\ &< 1 \text{ if } \quad nk > n_1 r. \end{aligned}$$

This implies that as n increases, the sequence $\{p_k(n)\}$ first increases and then decreases and reaches its maximum when n is the largest integer short of $n_1 r/k$. Therefore, $\hat{n}$ is the largest integer not exceeding $n_1 r/k$. When $n_1 = 88, r = 93, k = 19$, we have

$$\hat{n} = \left[\frac{n_1 r}{k}\right] = \left[\frac{88 \times 93}{19}\right] = 430.$$

Note that this would be the estimate of n if we assume that the proportion of marked *Rehu* fish and the unmarked *Rehu* fish caught on the second day are about the same.

Example 8: Some geneticists "believe" that a certain disease among human beings is inheritable—children inheriting the disease from their mothers. There is however no scientific basis for this belief. Suppose we wish to "test" the truth or otherwise of this belief based on observed data, how do we go about it? This is a typical problem of *testing of (statistical) hypothesis.* A (statistical) hypothesis is an assertion made about the unknown 'parameters'. In order to test the hypothesis that the disease is *not* heritable, data of the following type are generated: a sample of n children is classified according to (a) whether or not a child in the sample ever had the disease and, (b) whether or not the mother ever had the disease. These data can be displayed as below:

		Mother had the disease	Mother did not have the disease	Total
Child	Had the disease	a	b	$a+b$
	Did not have the disease	c	d	$c+d$
	Total	$a+c$	$b+d$	n

Now, if the proportion of children having the disease, given that their mothers also had the disease, (i.e., $a/(a+c)$) is significantly larger than the proportion of children having the disease, given that their mothers did not have the disease, (i.e., $b/(b+d)$), then we can conclude that these data provide sufficient evidence that the 'belief' of the geneticists is true. However, a question arises here: what is meant by "significantly larger"? To answer this question, we formalise the problem as below.

Based on the observed data, we wish to test the hypothesis that the fact that a child has or had the disease has nothing to do with whether or not his/her mother has or had the disease. Such an assertion is called a hypothesis of independence.

Suppose that the proportion *in the population* under consideration (namely all children born during a certain time period, assumed to be large) of children with disease born to mothers having the disease is α, the proportion of children with disease born to mothers not having the disease is β, the proportion of children not having disease born to mothers having the disease is ν and the proportion of children not having disease born to mothers not having the disease is δ.

The hypothesis can now be formulated as

$$\frac{\alpha}{\alpha+\nu} = \frac{\beta}{\beta+\delta}.$$

As stated earlier, if the hypothesis is true, then one would expect that $\frac{a}{a+c}$ is close to $\frac{b}{b+d}$ or equivalently, $(ad-bc)^2$ is small. A large observed value of $Q = (ad-bc)^2$ will indicate a departure from the hypothesis of independence. However, the term "large" is subjective and has to be properly quantified.

Suppose θ is a number such that

$$P(Q > \theta) = .05$$

if the hypothesis is true.

If the observed value of Q is bigger than θ, we argue that it is very unlikely (chance less than 5%) that the hypothesis of independence is true and we observe such a large value of Q.

So we conclude that the 'belief' that the diesease is inheritable is substantiated.

However, if observed $Q \leq \theta$, we conclude that it is possible to observe such a value even if the hypothesis is correct, and hence that data does not provide enough evidence to establish that the disease is heritable. For details on how to determine the value of θ, see [2].

The number .05 is called the level of significance, and is usually chosen to be .05 or .01. The number θ can be obtained from "chi-square" distribution tables which are commonly available.

The above examples demonstrate the nature of problems of statistical inference. A typical statistical inference problem can be formulated as follows: there is a population which is characterized by one or more fixed entities, called "parameters". For instance, in Example 7, n, the unknown number of *Rehu* fish in the pond is the parameter, while in Example 8, α, β, ν, and δ are the parameters. In practice, often some or all the parameters of the population will be unknown. Due to considerations of cost, time and/or space, it is often not possible (or, even desirable) to observe the entire population (even when the population is of finite size) and, what one can really observe in practice is a *sample* from the population. The size of the sample is usually not too large. Based on the sample observations, a statistician's job is to "infer" the unknown population parameters. Observe that in this process, the statistician is trying to extend from the particular (that is, sample data) to the general (that is, for the entire population).

Statistical inference is a hazardous process, because of the uncertainty present in the inference – one can never make perfect generalizations. However, uncertain inferences, based on sample data, can be made and the degree of uncertainty, as measured by the probability of making a wrong inference, can be specified, if the data on which inferences are based are collected according to some well defined principles. One of the key functions of statistical science is the provision of methods for making inferences and for measuring the degree of uncertainty of such inferences.

The problem of inferring on unknown parameters, based on sample data, has been tackled in two ways. The first one involves the consideration of certain function(s) of the sample observations to *guess* or *estimate* the unknown parameter(s). This is known as statistical estimation. Example 7 above illustrates one such situation. Alternatively, one can assign a hypothetical value to a parameter, that is, set up a statistical hypothesis and then test the truth of this hypothesis with a prescribed degree of confidence, based on the sample data. In Example 8 above, we have considered one such problem on testing.

6.6 Regression and Correlation

In many western countries, drunken driving is a serious offence. The law enforcement authorities have set a certain level of alcohol in the blood as acceptable and, a person having alcohol in his/her blood stream higher than the acceptable level is designated as "drunk". It is clear however that it is neither easy nor feasible to take blood samples of drivers and analyse them quickly to determine the alcohol level in their blood. Instead, what is done is to analyse the breath sample of a driver and determine the amount of a certain compound in the breath sample. This process is simple, quick and inexpensive. Based on this information, a driver can be classified as a drunk driver or not. Now the question arises: what does the amount of the compound measured from the breath sample have to do with the alcohol content in the blood? This can be answered effectively if one has evidence (based on large amount of observed data) that the alcohol content in the blood has an approximate linear relationship with the quantity of the compound determined from the breath sample. Indeed, it is now an established fact in medical science that the amount of alcohol present in the blood stream can be approximated by a linear function of the compound present in the breath sample. How did one arrive at such a fact? Initially, data on a large number of persons (of different age groups, body weight, etc.) were collected on both the variables, amount X of the compound in the breath sample and the corresponding alcohol content Y in the blood of the same person. Thus, if data on n persons were collected, then we have observations $(x_1, y_1), (x_2, y_2), \ldots, (x_n, y_n)$. If these pairs (treated as points in the X-Y plane) are plotted, then it is seen that these points are scattered around a straight line in the sense that some points are on the line, some are just below it and some others lie above but close to the line (see Figure 6.1).

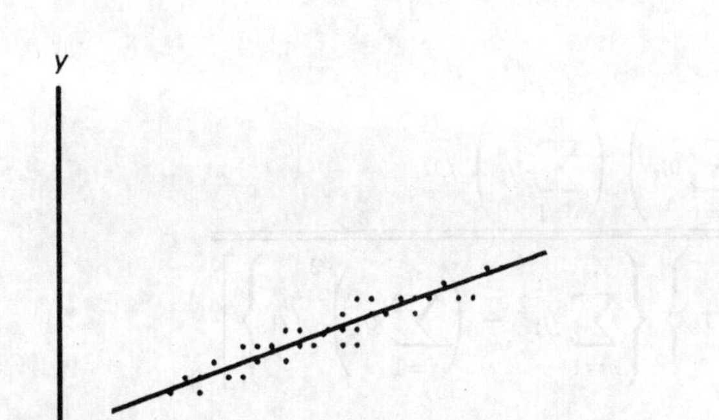

Figure 6.1

From this observation, it therefore seems reasonable to approximate the relationship between X and Y by a straight line, say $Y = a + bX$. Based on the data $\{(x_i, y_i);\ i = 1, 2, \ldots, n\}$, one can estimate the intercept a and the slope b of the line by $\hat{a}$ and $\hat{b}$ respectively, where $\hat{a}$ and $\hat{b}$ are given by

$$\hat{a} = \bar{y} - \hat{b}\bar{y};$$

$$\hat{b} = \frac{\sum_{i=1}^{n} x_i y_i - (\sum_{i=1}^{n} x_i)(\sum_{i=1}^{n} y_i)/n}{\sum_{i=1}^{n} x_i^2 - (\sum_{i=1}^{n} x_i)^2/n},$$

where $\bar{x} = \sum_{i=1}^{n} x_i/n$, $\bar{y} = \sum_{i=1}^{n} y_i/n$, are the sample means of X and Y respectively.

Thus, the 'fitted' straight line, obtained from the data, can be expressed as $y = \hat{a} + \hat{b}x$. The fitted line is often referred to as the (fitted) *regression line* of Y on X and $\hat{b}$ as the (estimated) *regression coefficient* of Y on X.

Now, if we wish to determine (or, more precisely, predict) the value of the alcohol content in the blood of a person for whom only the amount of the compound in the breath (as measured by the breath analyser), say x_0 is available, then we can use the fitted regression line to predict y for $x = x_0$ by simply substituting for x, the value x_0 in the regression equation $y = \hat{a} + \hat{b}x$ and computing the corresponding value of y. The computed value of y, say y_0 as obtained from the regression equation, (i.e., $y_0 = \hat{a} + \hat{b}x_0$) is the predicted value of y for $x = x_0$. Note that since the relationship between X and Y has been approximated by a straight line, it is not necessary that all values of Y for a given value of X will lie on the fitted line.

Another question that arises is: how good is the fitted straight line as an approximation to the 'true' relationship between X and Y? To measure the goodness of the fitted line, one can compute the *correlation coefficient* between X and Y from the data, using the formula

$$r = \frac{\sum_{i=1}^{n} x_i y_i - \left(\sum_{i=1}^{n} x_i\right)\left(\sum_{i=1}^{n} y_i\right)/n}{\sqrt{\left[\left\{\sum_{i=1}^{n} x_i^2 - \left(\sum_{i=1}^{n} x_i\right)^2/n\right\}\left\{\sum_{i=1}^{n} y_i^2 - \left(\sum_{i=1}^{n} y_i\right)^2/n\right\}\right]}}$$

$$= \frac{\sum_{i=1}^{n} (x_i - \bar{x})(y_i - \bar{y})}{\left[\left\{\sum_{i=1}^{n} (x_i - \bar{x})^2\right\}\left\{\sum_{i=1}^{n} (y_i - \bar{y})^2\right\}\right]^{1/2}},$$

where, as before, $\bar{x}$ and $\bar{y}$ stand for the sample means of X and Y respectively.

Note that r, as defined above, is dimensionless. Also, $r^2 \leq 1$ (or, equivalently, $-1 \leq r \leq 1$)). This is a consequence of the well known Cauchy inequality: if $u_1, u_2, \ldots, u_n$ and $v_1, v_2, \ldots, v_n$ be two sets of real numbers, then the Cauchy inequality states that

$$\left(\sum_{i=1}^{n} u_i v_i\right)^2 \leq \left(\sum_{i=1}^{n} {u_i}^2\right)\left(\sum_{i=1}^{n} {v_i}^2\right)$$

To see the truth of this inequality, observe that for any real number λ,

$$\sum_{i=1}^{n}(u_i - \lambda v_i)^2 \geq 0,$$

or,

$$\left(\lambda\sqrt{\sum {v_i}^2} - \frac{\sum u_i v_i}{\sqrt{\sum {v_i}^2}}\right)^2 + \sum {u_i}^2 - \frac{\left(\sum u_i v_i\right)^2}{\sum {v_i}^2} \geq 0, \text{ for all real } \lambda.$$

In particular, the above inequality holds for

$$\lambda = \sum u_i v_i / \sum {v_i}^2.$$

Plugging in this value of λ in the last inequality, we have

$$\sum {u_i}^2 - \frac{(\sum u_i v_i)^2}{\sum {v_i}^2} \geq 0,$$

which is the Cauchy inequality. The equality holds if and only if for some constant α, $v_i = \alpha u_i$ for all i (**Exercise**).

How do we interpret an observed value of r? If $r = \pm 1$, then there exists a perfect linear relationship between X and Y (**Exercise**). Thus, a value of r close to ± 1 indicates that the strategy of approximating the relationship between X and Y by a straight line is indeed a sound one. On the other hand, low values of r (say $r < .5$) indicates that the approximation of the relationship between X and Y by a straight line is a poor one and that further investigation into the relationship is necessary.

So far, we have talked about linear regression, (i.e., where the relationship between X and Y has been approximated by a straight line). In practice, often one encounters situations where the plot of the data indicates that X and Y may not be linearly related. One such situation, for instance, is in studying the effect of a plant nutrient (say, Nitrogen) on the yield of a crop. At lower doses, the yield is seen to increase in a linear fashion, but after a certain level of the nutrient, the increase in the yield for a unit change in the amount of nutrient starts decreasing. In such situations, one needs to approximate the relationship between X and Y by some function other than the linear, say, by a parabola. We, however, do not elaborate on this point as it is beyond the scope of this book.

Suggested Reading

1. *An Introduction to Probability Theory and its Applications* by W.Feller, Wiley Eastern, New Delhi, 1978 (read Chapters 1–7).

2. *Introduction to Mathematical Statistics* by R. V. Hogg and A. T. Craig, Macmillan, New York, 1978.

Chapter 7

Algebraic Structures and Geometry

"Symmetry, as wide or as narrow you may define, is one idea by which man through the ages has tried to comprehend and create order, beauty and perfection."

Herman Weyl

7.1 Groups

All of us have some notion of symmetry. In fact, we draw many geometric conclusions by looking at symmetries of figures. In daily experience also, we come across many kinds of symmetries. Let us look at a few of them (Figure 7.1). A circle is one of the earliest symmetrical figures we come across. Similarly an equilateral triangle is a symmetrical figure, so is an isosceles triangle, whereas a triangle which is not isosceles or equilateral is unsymmetrical. In the case of a circle, if we fix its center and rotate it through any angle, the figure coincides with itself.

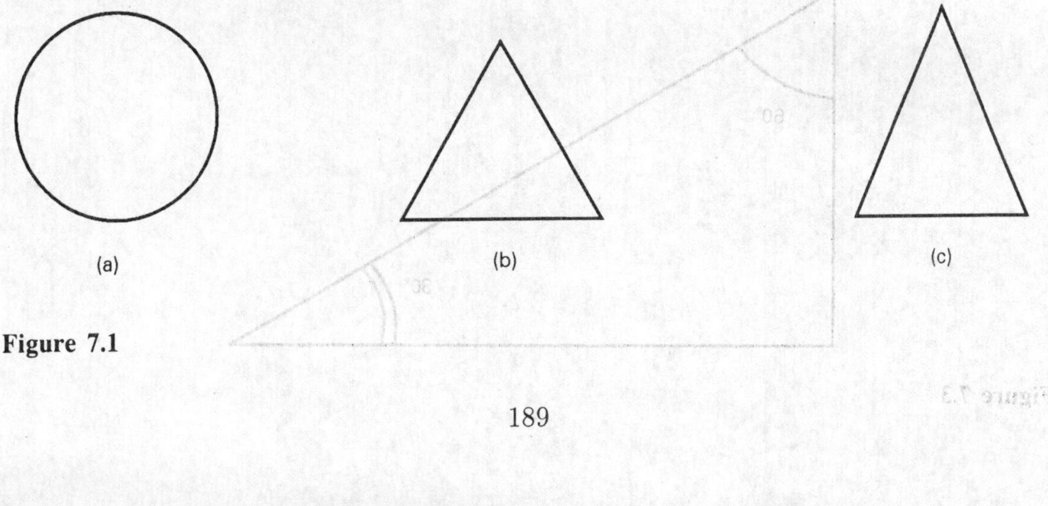

Figure 7.1

Similarly if we reflect the circle about any of its diameters then it coincides with itself. In the case of an equilateral triangle and an isosceles triangle, it is obvious that an equilateral triangle is *more* symmetrical than an isosceles triangle which is not equilateral. Now what do we mean by 'more symmetrical'?

If we reflect an equilateral triangle about any of its medians or if we rotate the figure fixing its centroid through 120°, then it coincides with itself. On the other hand, the only symmetry an isosceles triangle which is not equilateral has, is reflection about the median which passes through the midpoint of one of its unequal sides and the opposite vertex. No rotation brings it back to its own position, save the trivial rotation through 360°. So an equilateral triangle has more symmetries than an isosceles tringle which is not equilateral.

One can ask many similar questions. How many symmetries do a rhombus, rectangle and a square have (Figure 7.2)?

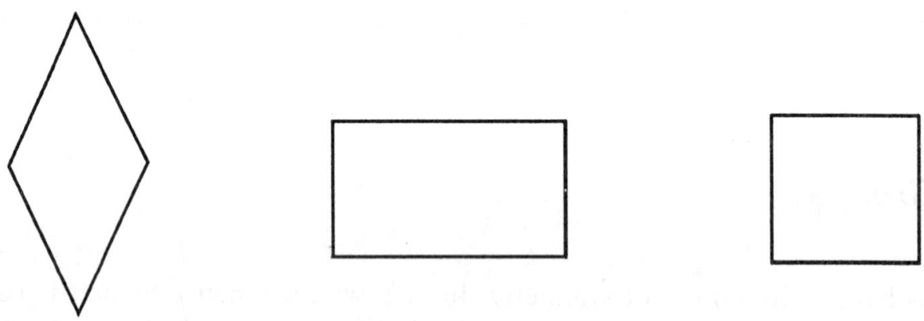

Figure 7.2

What about a triangle whose angles are 30°, 60° and 90° respectively (Figure 7.3)? **(Exercise)**

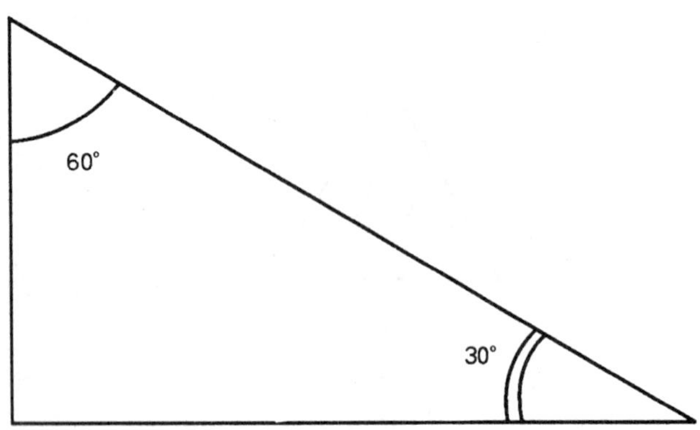

Figure 7.3

Also, what are the symmetries of an infinite row of triangles arranged as below (Figure 7.4)?

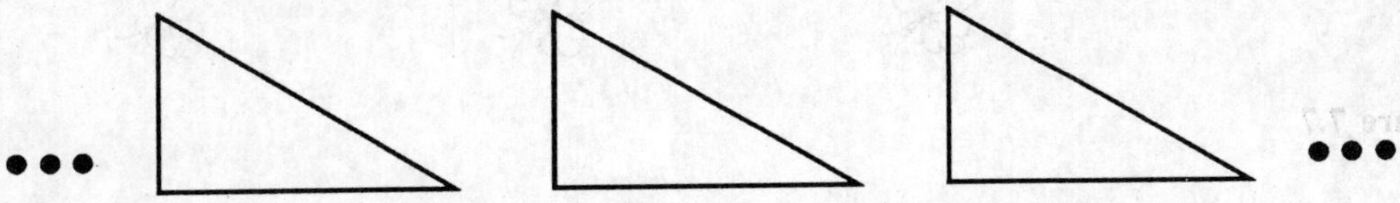

Figure 7.4

Surely we cannot call it unsymmetrical. If we "translate" it along a line through a fixed distance, it is going to coincide with itself. Similarly, the two-dimensional lattice in Figure 7.5 remains the same by a horizontal as well as a vertical translation by unit lengths.

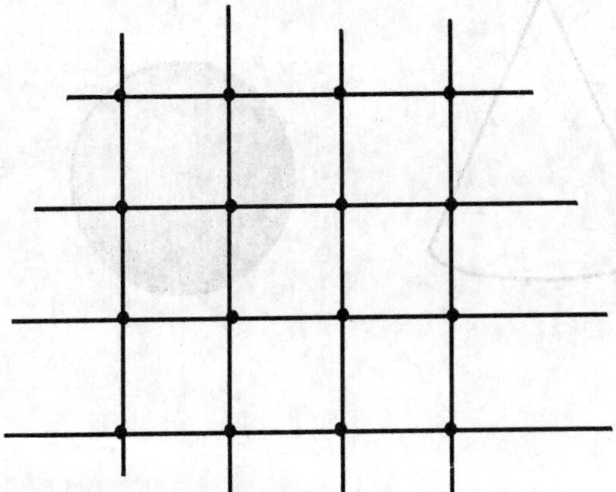

Figure 7.5

Look at Figures 7.6 and 7.7.

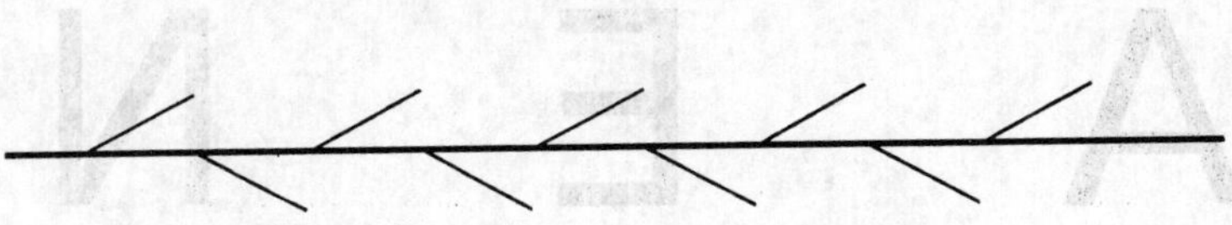

Figure 7.6

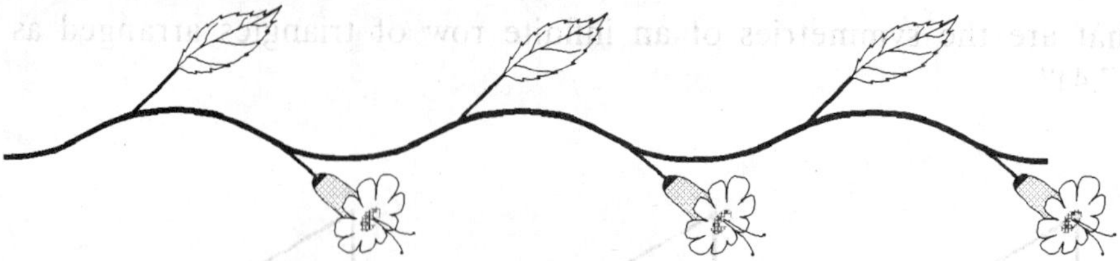

Figure 7.7

What symmetries do they have? While the first picture remains the same after a translation and a reflection about the horizontal line, the second remains so after a translation only.

If we now move to three-dimensional space, then a cube and a regular tetrahedron are a few examples of symmetrical objects. Of course, a sphere is an object with highest symmetry. A right circular cone (Figure 7.8) is also very symmetrical.

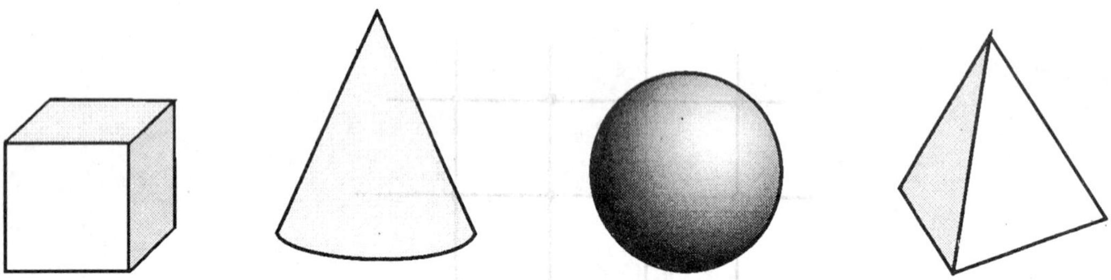

Figure 7.8

The sphere, of course, comes back to itself when we rotate it by fixing its center, whereas a cone comes back to itself when we rotate it through any angle about its axis.

It is interesting also to look at the figures (Figures 7.9–7.12) and to find the kind of symmetries that they have.

Figure 7.9

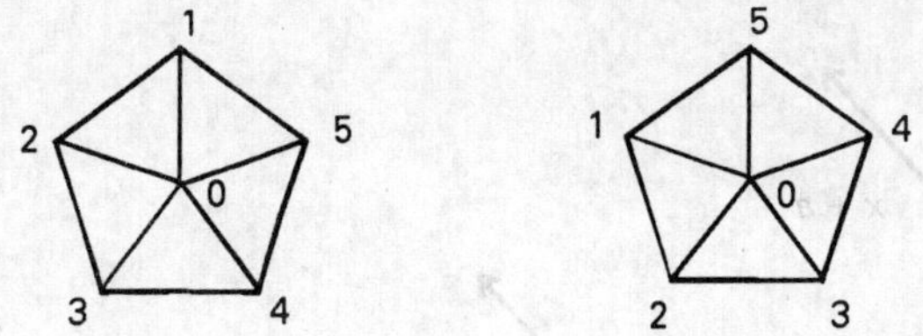

Figure 7.10

Figure 7.11

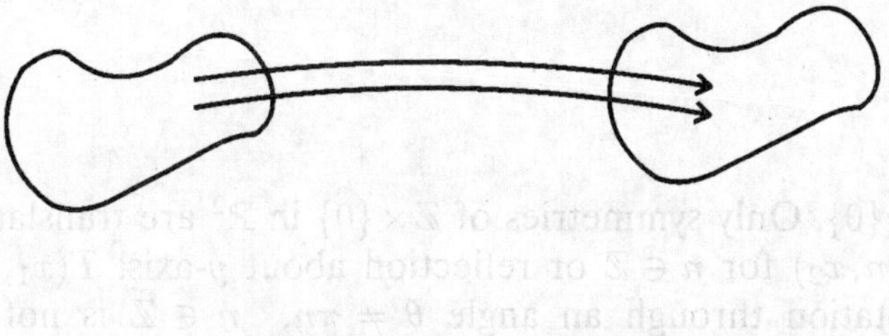

Figure 7.12

Next we try to understand what symmetries are.

DEFINITION: As it must be clear to the reader by now, by symmetry of a figure $A \subset \mathbb{R}^2$ we mean a transformation $T: \mathbb{R}^2 \longrightarrow \mathbb{R}^2$ such that $\|Tx - Ty\| = \|x - y\|$ and $T(A) = A$, where $T(A)$ stands for the range of A under T and $\| u \|$ stands for the length of the vector $u \in \mathbb{R}^2$ (for $u = (u_1, u_2)$, $\| u \| = \sqrt{u_1^2 + u_2^2}$; see Chapter 5).

Moreover, a transformation $T: \mathbb{R}^2 \longrightarrow \mathbb{R}^2$ is called an *isometry* if $\|T(x) - T(y)\| = \|x - y\|$ for all $x, y \in \mathbb{R}^2$, meaning T preserves distance between any two points of $\mathbb{R}^2$. Thus a symmetry of $A \subseteq \mathbb{R}^2$ is an isometry T in $\mathbb{R}^2$ such that $T(A) = A$. Let us look at a few examples of isometry.

TRANSLATION For a fixed $a \in \mathbb{R}^2$, let $T: \mathbb{R}^2 \longrightarrow \mathbb{R}^2$ be defined by $T_a(x) = x + a$. The action of T is to translate every point x parallel to the given vector a by an amount equal to the length of a (Figure 7.13).

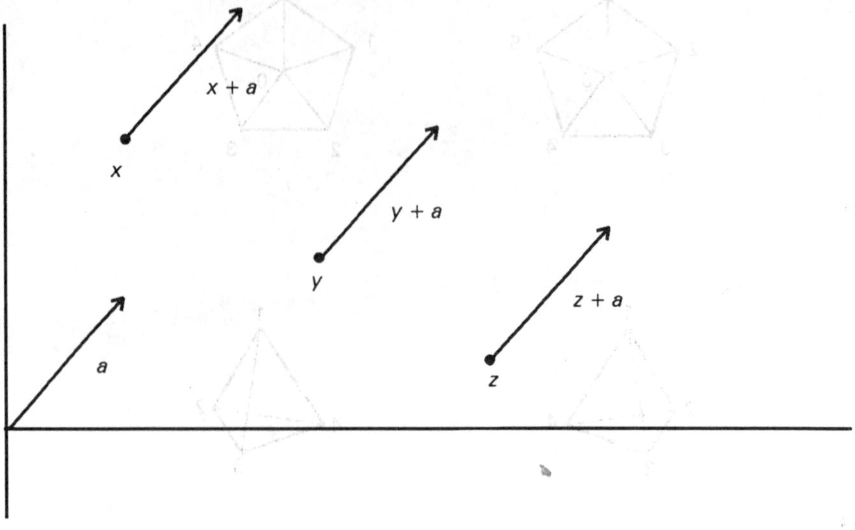

Figure 7.13

Consider $A = \mathbb{Z} \times \{0\}$. Only symmetries of $\mathbb{Z} \times \{0\}$ in $\mathbb{R}^2$ are translations of the type $T(x_1, x_2) = (x_1 + n, x_2)$ for $n \in \mathbb{Z}$ or reflection about y-axis: $T(x_1, x_2) = (-x_1, x_2)$. Note that any rotation through an angle $\theta \neq \pi n, \quad n \in \mathbb{Z}$ is not a symmetry of $\mathbb{Z} \times \{0\}$ though it is an isometry.

Rotation: For a fixed $\theta \in [-\pi, \pi)$, let $R_\theta x = (x_1 \cos\theta - x_2 \sin\theta, \;\; x_1 \sin\theta + x_2 \cos\theta)$, where $x = (x_1, x_2)$. This obviously is a rotation in the $\mathbb{R}^2$-plane, keeping the origin fixed.

Reflection: Consider, for example, the transformation $S(x) = (x_1, -x_2)$. Clearly, S is an isometry. But there is no angle θ such that one can write $S(x) = R_\theta(x)$, i.e., such a transformation S cannot be obtained as an appropriate rotation. In fact, S is a reflection of the $\mathbb{R}^2$-plane about the X-axis. In general, one can get a reflection in $\mathbb{R}^2$ about any line in $\mathbb{R}^2$ (Figure 7.14).

If we want to reflect the point (x_1, x_2) on the line $ax + by + c = 0$, we get the point

$$\left(\frac{(b^2 - a^2)x_1 - 2abx_2 - 2ca}{a^2 + b^2}, \;\; \frac{-2abx_1 + (a^2 - b^2)x_2 - 2bc}{a^2 + b^2}\right)$$

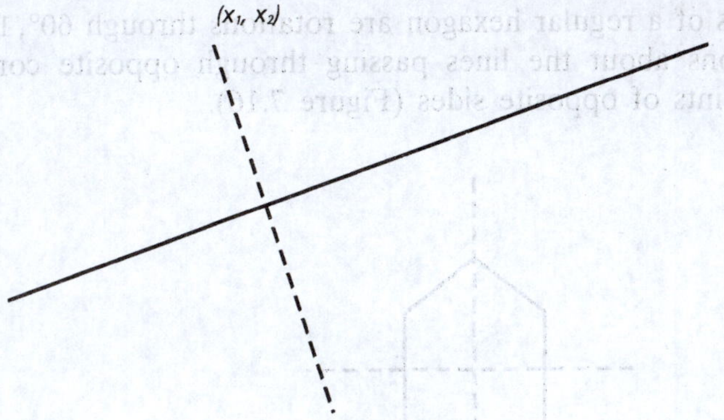

Figure 7.14

Exercise: Derive this formula and show that it is an isometry and cannot be obtained as a rotation.

We observe that the composition of two isometries is an isometry. It is convenient to write (x_1, x_2) as $\begin{pmatrix} x_1 \\ x_2 \end{pmatrix}$, so that

$$R_\theta \begin{pmatrix} x_1 \\ x_2 \end{pmatrix} \equiv \begin{pmatrix} x_1 \cos\theta - x_2 \sin\theta \\ x_1 \sin\theta + x_2 \cos\theta \end{pmatrix} = \begin{pmatrix} \cos\theta & -\sin\theta \\ \sin\theta & \cos\theta \end{pmatrix} \begin{pmatrix} x_1 \\ x_2 \end{pmatrix}.$$

Thus R_θ is identified with the matrix $\begin{pmatrix} \cos\theta & -\sin\theta \\ \sin\theta & \cos\theta \end{pmatrix}$.

For an equilateral triangle $\Delta \subset \mathbb{R}^2$ with its centroid at the origin, rotation through 120° and 240° are symmetries of Δ. A reflection about a line passing through one of its vertices and the midpoint of its opposite side is also a symmetry of Δ.

Similarly, the symmetries of a square with its centre at the origin are rotation through $90°, 180°, 270°$, and reflections about the lines passing through the mid points of the opposite sides and reflection on its diagonals.

The symmetries of a regular pentagon are rotations of $72°, 144°, 216°, 288°$ about its center and reflection about the lines passing through the vertices and midpoints of opposite sides (Figure 7.15),

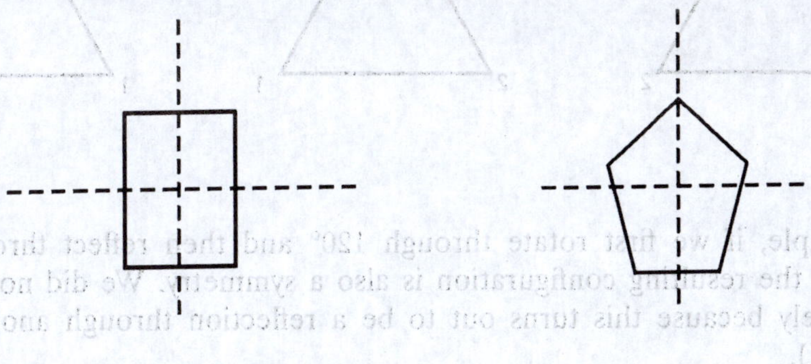

Figure 7.15

whereas the symmetries of a regular hexagon are rotations through 60°, 120°, 180°, 240°, 300° and reflections about the lines passing through opposite corners and lines joining the midpoints of opposite sides (Figure 7.16).

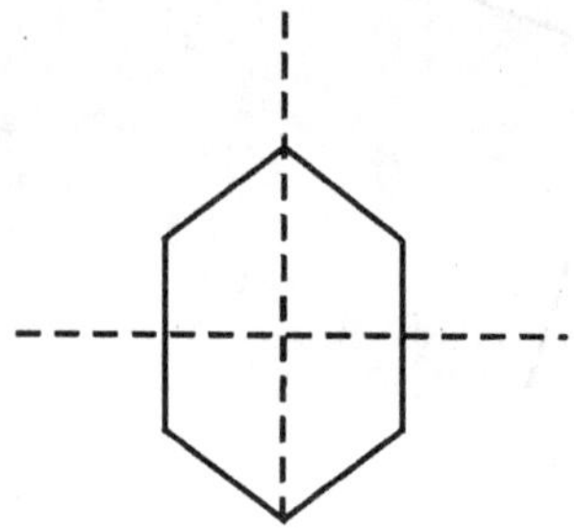

Figure 7.16

As we have seen, there are five symmetries of an equilateral triangle △ if we do not consider the trivial identity mapping. One could compose two symmetries of △ and the map thus obtained is also a symmetry (Figure 7.17).

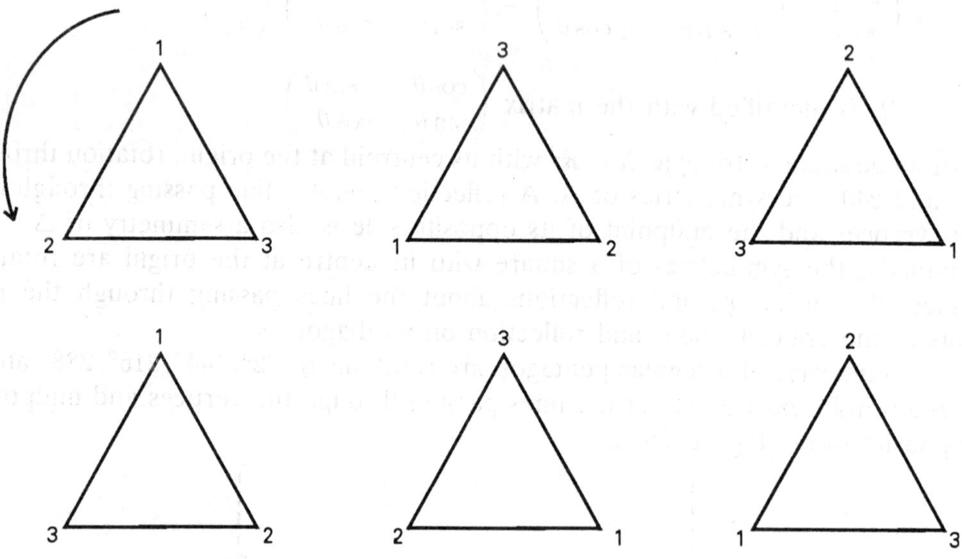

Figure 7.17

For example, if we first rotate through 120° and then reflect through one of its medians, the resulting configuration is also a symmetry. We did not enumerate this seperately because this turns out to be a reflection through another median (Figure 7.17).

Like isometry in $\mathbb{R}^2$, we may define an isometry in $\mathbb{R}^3$ as a map $T : \mathbb{R}^3 \longrightarrow \mathbb{R}^3$ such that $\| T(x) \| = \| x \|$, where $\| x \| = \sqrt{x_1^2 + x_2^2 + x_3^2}$, for $x = (x_1, x_2, x_3)$. Examples of isometry here too are translation, rotation and reflection. Translation $T_a : \mathbb{R}^3 \longrightarrow \mathbb{R}^3$ is easily defined by $T_a(x) = x + a$. In analogy with $\mathbb{R}^2$, an improper rotation (composed of rotations and reflections) in $\mathbb{R}^3$ can be given as a transformation: $R(x) = (l_1x_1 + m_1x_2 + n_1x_3, l_2x_1 + m_2x_2 + n_2x_3, l_3x_1 + m_3x_2 + n_3x_3)$, where $l_1^2 + m_1^2 + n_1^2 = l_2^2 + m_2^2 + n_2^2 = l_3^2 + m_3^2 + n_3^2 = 1$, $l_1l_2 + m_1m_2 + n_1n_2 = l_1l_3 + m_1m_3 + n_1n_3 = l_3l_2 + m_3m_2 + n_3n_2 = 0$ and also $l_1^2 + l_2^2 + l_3^2 = m_1^2 + m_2^2 + m_3^2 = n_1^2 + n_2^2 + n_3^2 = 1$, $l_1m_1 + l_2m_2 + l_3m_3 = l_1n_1 + l_2n_2 + l_3n_3 = m_1n_1 + m_2n_2 + m_3n_3 = 0$. Represented as a matrix it looks like

$$R\begin{pmatrix} x_1 \\ x_2 \\ x_3 \end{pmatrix} = \begin{pmatrix} l_1 & m_1 & n_1 \\ l_2 & m_2 & n_2 \\ l_3 & m_3 & n_3 \end{pmatrix}\begin{pmatrix} x_1 \\ x_2 \\ x_3 \end{pmatrix}.$$

A simple computation using matrix multiplication rules (Chapter 5) and the above relations leads to $RR' = R'R = I$, where R' is the transpose matrix of R and I is the 3×3 identity matrix (**Exercise**). Note that this definition includes reflections, for example, in the above notation for R, take $l_1 = -m_2 = n_3 = 1$ and the rest of the entries zero. It can be verified (**Exercise**) that (i)$RR' = R'R = I$, (ii) $det(R) = -1$, (iii) R maps (x_1, x_2, x_3) to $(x_1, -x_2, x_3)$.

In contrast to $\mathbb{R}^2$, here a reflection shall mean a reflection in a plane. The reader is encouraged to find a general formula for reflection in $\mathbb{R}^3$ (**Exercise**).

A little thought should convince the reader that translation, rotation or composition of both not only preserves the distance but also the "orientation", but a reflection *does not preserve orientation* even when it preserves the distance. This is experienced when one looks at one's image in a plane mirror; one sees it very much like oneself but the left hand becomes the right hand of the image and vice-versa, i.e., the sense of left-right gets interchanged.

Recall that a symmetry transformation of $A \subset \mathbb{R}^3$ is an isometry $T : \mathbb{R}^3 \longrightarrow \mathbb{R}^3$ such that $T(A) = A$. Now given a set A, consider the set $G = \{T : T \text{ is a symmetry of the set } A\}$. We see that the composition of two symmetries of A is also a symmetry of A. So if $T_1, T_2 \in G$, then $T_1 \circ T_2 \in G$, where $T_1 \circ T_2$ stands for composition of T_1 and T_2. We observe further that

(1) For $T_1, T_2, T_3 \in G$, $T_1 \circ (T_2 \circ T_3) = (T_1 \circ T_2) \circ T_3$.

(2) There is $I \in G$ such that $T \circ I = I \circ T = T$, for all $T \in G$, where I is the identity transformation (which means no transformation at all).

(3) For all $T \in G$ there is a unique element $S \in G$ such that $T \circ S = S \circ T = I$.

The property (3) says that there is a unique symmetry transformation S which undoes the effect of a given symmetry transformation T and this follows from the fact that T is one-to-one and onto (**Exercise**). We shall write $S = T^{-1}$.

A set G, equipped with a binary operation $\star$, is a *group* if the following three properties are satisfied:

(1)′ $a \star (b \star c) = (a \star b) \star c$, for all $a, b, c \in G$,

(2)′ there is an $e \in G$ such that $e \star a = a \star e$, for all $a \in G$ and

(3)′ for every $a \in G$, there is a unique $b \in G$ such that $a \star b = b \star a = e$.

The element e is called the identity of the group and the unique element b in (3′) the *inverse* of a, written a^{-1}. Thus, we immediately recognise that the set of symmetries G of a set $A \subseteq \mathbb{R}^2$ or $\mathbb{R}^3$ is one such group and G is called the *group of symmetries* of A.

The identity and the two elements of the group of symmetries corresponding to the rotations of an equilateral triangle with its centre at the origin in matrix representation, are given as

$$\begin{pmatrix} 1 & 0 \\ 0 & 1 \end{pmatrix}, \quad \begin{pmatrix} -\frac{1}{2} & -\frac{\sqrt{3}}{2} \\ \frac{\sqrt{3}}{2} & -\frac{1}{2} \end{pmatrix}, \quad \begin{pmatrix} -\frac{1}{2} & \frac{\sqrt{3}}{2} \\ -\frac{\sqrt{3}}{2} & -\frac{1}{2} \end{pmatrix}.$$

If we take the triangle to be symmetrical about the y-axis, then the reflection about the y-axis shall correspond to the matrix

$$\begin{pmatrix} -1 & 0 \\ 0 & 1 \end{pmatrix}.$$

Since they are elements of a group, we see that

$$\begin{pmatrix} -1 & 0 \\ 0 & 1 \end{pmatrix} \begin{pmatrix} -\frac{1}{2} & +\frac{\sqrt{3}}{2} \\ -\frac{\sqrt{3}}{2} & -\frac{1}{2} \end{pmatrix} = \begin{pmatrix} \frac{1}{2} & -\frac{\sqrt{3}}{2} \\ -\frac{\sqrt{3}}{2} & -\frac{1}{2} \end{pmatrix}$$

is also a symmetry of the triangle. This is nothing but the reflection about one of the other medians.

So the group of symmetries of an equilataeral triangle shall correspond to the group of matrices

$$\left\{\begin{pmatrix}1&0\\0&1\end{pmatrix},\begin{pmatrix}-\frac{1}{2}&-\frac{\sqrt{3}}{2}\\\frac{\sqrt{3}}{2}&-\frac{1}{2}\end{pmatrix},\begin{pmatrix}-\frac{1}{2}&\frac{\sqrt{3}}{2}\\-\frac{\sqrt{3}}{2}&-\frac{1}{2}\end{pmatrix},\begin{pmatrix}\frac{1}{2}&\frac{\sqrt{3}}{2}\\\frac{\sqrt{3}}{2}&-\frac{1}{2}\end{pmatrix},\begin{pmatrix}\frac{1}{2}&-\frac{\sqrt{3}}{2}\\-\frac{\sqrt{3}}{2}&-\frac{1}{2}\end{pmatrix},\begin{pmatrix}-1&0\\0&1\end{pmatrix}\right\}$$

If we write

$$I=\begin{pmatrix}1&0\\0&1\end{pmatrix},\ T=\begin{pmatrix}-\dfrac{1}{2}&-\dfrac{\sqrt{3}}{2}\\\dfrac{\sqrt{3}}{2}&-\dfrac{1}{2}\end{pmatrix},$$

and

$$S=\begin{pmatrix}-1&0\\0&1\end{pmatrix},$$

then

$$T^2=\begin{pmatrix}-\dfrac{1}{2}&\dfrac{\sqrt{3}}{2}\\-\dfrac{\sqrt{3}}{2}&-\dfrac{1}{2}\end{pmatrix},\quad ST=\begin{pmatrix}\dfrac{1}{2}&\dfrac{\sqrt{3}}{2}\\\dfrac{\sqrt{3}}{2}&-\dfrac{1}{2}\end{pmatrix},\quad ST^2=\begin{pmatrix}\dfrac{1}{2}&-\dfrac{\sqrt{3}}{2}\\-\dfrac{\sqrt{3}}{2}&-\dfrac{1}{2}\end{pmatrix}$$

and

$$T^3=I.$$

Thus the elements of the group can also be denoted by $\{I,T,T^2,S,ST,ST^2\}$. Observe further that $S^2=I,(ST)^2=I,(ST^2)^2=I$. In a way the whole group is *generated* by just two elements T and S. We shall return to this later. (**Exercise**: Compute the group of symmetries of a regular hexagon in matrix form).

What about the symmetry of a circle? It is easy to see that there are infinitely many symmetries of the circle, i.e., rotation through every angle θ is a symmetry of the circle. Similarly $\mathbb{Z}\times\{0\}$ has infinitely many symmetries, though the only symmetries are translations by integral units parallel to the x-axis. The two-dimensional lattice $\mathbb{Z}\times\mathbb{Z}$ has also infinitely many symmetries consisting of translations by integral units, parallel to the x-axis and y-axis.

We have defined a group as an abstract entity of which the sets of symmetries of figures are examples. Consider $\mathbb{Z}$ under natural addition (+), and observe that it is truly a group with 0 as identity, since

1. $m+(n+p)=(m+n)+p,\quad$ for all $\quad m,n,p\in\mathbb{Z},$
2. $m+0=0+m=m,\quad$ for all $\quad m\in\mathbb{Z},$

3. $m+(-m)=0, \quad \text{for all} \quad m \in \mathbb{Z}.$

Similarly, $\mathbb{R}_{+0}$, the set of non-zero positive real numbers, is a group under the usual multiplication:

1. $x \cdot (y \cdot z) = (x \cdot y) \cdot z, \quad \forall \quad x, y, z \in \mathbb{R}_+ \quad ,$
2. $x \cdot 1 = 1 \cdot x = x, \quad \forall x \in \mathbb{R}_+ \quad ,$
3. $x \cdot x^{-1} = x^{-1} \cdot x = 1, \quad \forall \quad x \in \mathbb{R}_+ \quad .$

Here, the identity element is 1 and the inverse of a positive number x is its reciprocal. (**Exercise**: Verify that the set of real numbers $\mathbb{R}$ is a group under usual addition (+) but not a group under the usual multiplication, while $\mathbb{R} \setminus \{0\}$, the set of all non-zero real numbers, is a group under multiplication.)

The sets

$$Gl(n, \mathbb{R}) = \{T : \ T \quad n \times n \text{ invertible matrix with real entries}\},$$

$$O(n) = \{T : \quad T \in Gl(n, \mathbb{R}) \quad \text{such that } T'T = TT' = I\}$$

for $n = 2, 3$, are examples of groups. (Here T' stands for the transpose of T). $O(n)$ is called the orthogonal group in n dimensions. The reason for this nomenclature is that the rows of the matrix are mutually orthogonal, as also the columns (see Chapter 5). We observe that the elements of the orthogonal group preserve length in $\mathbb{R}^3$ and hence, intuitively speaking, must be a rotation. For $n = 2$, we easily see that this implies that det $T = \pm 1$. In case $\det T = 1$, then A not only preserves length but also orientation. In fact an element of $O(3)$ which preserves orientation is called a *proper rotation*, and such elements also form a group denoted by $SO(3)$. The above ideas clearly extend to any natural number n.

Permutations: While looking at symmetries of an equilateral triangle ABC, it was enough to look at what happened to the vertices A, B and C. A rotation through $\frac{2\pi}{3}$ radians caused the vertex A to move to the position which B was occupying, whereas B moved to the position which C was occupying and C to the position where A was. Thus it appears that a rotation of the triangle ABC is a rearrangement of the vertices A, B and C. Recall what is meant by a permutation of a set of elements. For a set E we call a bijection $f: \quad E \longrightarrow E$ a permutation of the elements of the set E. For example a map

$$f: \ \{A, B, C\} \longrightarrow \{A, B, C\}$$

such that $f(A) = B, f(B) = C, f(C) = A$ is precisely the rotational symmetry transformations of an equilateral triangle ABC. This can sometimes be represented as $A \mapsto B, B \mapsto C, C \mapsto A$ or $\begin{pmatrix} A & B & C \\ B & C & A \end{pmatrix}$.

It is not difficult to see that the set of all permutations of a given set E, forms a group $S(E)$ under composition of permutations. If E has n elements then $S(E)$ has $n!$ elements.

The reader should have noted that the set of only rotational transformations of an equilateral is a group in its own right.

Multiplication Tables: It is clear that given a group G with binary operation $\star$, the group is known if we know the composition $a \star b$ of every pair of elements $a, b \in G$. So if a group G has elements $\{e, a, b, c, \cdots\}$, where e is the identity element, then we ought to know the table.

$\star$	e	a	b	c	$\cdot$
e	e	a	b	c	$\cdot$
a	a	$a \star a$	$a \star b$	$a \star c$	$\cdot$
b	b	$b \star a$	$b \star b$	$b \star c$	$\cdot$
c	c	$c \star a$	$c \star b$	$c \star c$	$\cdot$
$\cdot$	$\cdot$	$\cdot$	$\cdot$	$\cdot$	$\cdot$

For example, the group of the rotational symmetries of an equilateral triangle as was shown earlier is given by $\{I, T, T^2\}$, which has the multiplication table given below:

$\star$	I	T	T^2
I	I	T	T^2
T	T	T^2	I
T^2	T^2	I	T^1

The group of all symmetries of an equilateral triangle as was shown earlier is given by $\{I, T, T^2, S, ST, ST^2\}$ with the multiplication table

	I	T	T^2	S	ST	ST^2
I	I	T	T^2	S	ST	ST^2
T	T	T^2	I	ST^2	S	ST
T^2	T^2	I	T	ST	ST^2	S
S	S	ST	ST^2	I	T	T^2
ST	ST	ST^2	S	T^2	I	T
ST^2	ST^2	S	ST	T	T^2	I

(**Exercise**: Verify this table).

Consider a set $\{e, a, b, c\}$, with the multiplication table:

	e	a	b	c
e	e	a	b	c
a	a	b	c	e
b	b	c	e	a
c	c	e	a	b

It is obvious from the table that under the operation defined by the table, $\{e, a, b, c\}$ constitutes a group. If instead we have the same set with the modified multiplication table

	e	a	b	c
e	e	a	b	c
a	a	e	c	b
b	b	c	e	a
c	c	b	a	e

then this is also a group of four elements but with a different rule of composition. The reader might think that by renaming a, b, c one may get a rule of composition similar to the above. However, one can convince oneself that this is not possible. To understand this, we proceed as follows.

Suppose G_1, G_2 are two groups. A map $\varphi : \ G_1 \longrightarrow G_2 \quad (G_1, G_2$ considered as sets) is called a *homomorphism* if it respects the group structure. This means that $\varphi(a \star b) = \varphi(a) \circ \varphi(b)$, where $\star$ is the binary operation in G_1 and $\circ$ is the binary operation in G_2, $\varphi(e_1) = e_2$, where e_1 and e_2 are identities of G_1 and G_2 respectively and $\varphi(a^{-1}) = \varphi(a)^{-1}$. For example, if we define $\varphi : \ \mathbb{R} \longrightarrow \mathbb{R}_+$ by $\varphi(x) = 2^x$, one sees that $\varphi(x+y) = 2^{x+y} = 2^x \cdot 2^y = \varphi(x)\varphi(y)$ ($\varphi(0) = 1$ and $\varphi(-x) = 2^{-x} = \varphi(x)^{-1}$). Thus $\mathbb{R}$ as a group under the operation $+$ is mapped by φ onto a group under the operation of multiplication. Similarly, logarithm is a homomorphism from $\mathbb{R}_+$ to $\mathbb{R}$. (**Exercise**: Verify $\varphi : \mathbb{R} \to O(2)$ defined by $\varphi(t) = \begin{pmatrix} \cos t & -\sin t \\ \sin t & \cos t \end{pmatrix}$ is a homomorphism.)

We call a bijective homomorphism between two groups an *isomorphism*. (Note that the inverse of a bijective homomorphism is also a homomorphism). If there is an isomorphism between two groups then they are called isomorphic. With regard to the two groups of four elements that were described above, we now ask: are they isomorphic? The answer is clearly negative because while there are three elements a, b, c in the first group such that $a \star a = b \star b = c \star c = e$, in the other group there is only one element with such a relation. An element of the group G is said to be of order $n \in \mathbb{N}$ if $a^n = e$, but $a^k \neq e$ for $1 \leq k < n$. For example, in the second multiplication table on page 201, the element b is of order 2 whereas c is of order 4. On the other hand in the third table, every element other than the identity is of order 2.

Subgroups: A subset H of a group G is called a *subgroup* of G if H is a group with the operation of the group G. Thus for $H \subset G$, H is a subgroup if

(i) $h_1, h_2 \in H$ implies that $h_1 h_2 \in H$,
(ii) $h \in H$ implies that $h^{-1} \in H$, and
(iii) $e \in H$.

Examples

(i) $\mathbb{Z}$ is a subgroup of $\mathbb{R}$ under addition.

(ii) If the group of all symmetries of an equilateral triangle is denoted by G, then the group of rotations through 120°, 240° and 360° fixing its centroid is a subgroup of G.

(iii) $O(n)$ is a subgroup of $Gl(n, \mathbb{R})$.

(iv) If S_n is the group of permutatians in n symbols $\{1, 2, \cdots, n\}$, then S_m is a subgroup of S_n for $m < n$.

(v) Define $Sl(2, \mathbb{R}) = \{T \in Gl(2, \mathbb{R}) : \ \det T = 1\}$. $Sl(2, \mathbb{R})$ is a subgroup of $Gl(2, \mathbb{R})$.

(**Exercise**: Verify these statements)

A group G is said to be of *finite order* if it has finitely many elements and the number of elements of G is called the *order* of G, and if it is not of finite order then G is said to be an *infinite* group. In the above set of examples, (ii), (iv) are that of finite groups while the rest are of infinite ones. We shall demonstrate a very important result (due to Lagrange) of group theory which says that the order of a subgroup of a finite group divides the order of the group.

Let G be a group with H as a subgroup. Let us define a relation $\sim$ in G by $a \sim b$ if $ab^{-1} \in H$. It is clear that $a \cdot a^{-1} = e \in H$ so that $a \sim a$ for every $a \in G$. If $a \sim b$, then $ab^{-1} \in H$ and therefore $ba^{-1} = (ab^{-1})^{-1} \in H$, making $b \sim a$. If $a \sim b$ and $b \sim c$, then ab^{-1} and bc^{-1} are in H and hence $ac^{-1} = (ab^{-1})(bc^{-1}) \in H$. Thus we have verified the reflexivity, symmetry and transitivity of $\sim$ making it an equivalence relation in G, and therefore it partitions G into disjoint equivalence classes (see Section 1.5). What do these equivalence classes in G look like?

The equivalence class of an element $a \in G$ consists of

$$\{b : \ ab^{-1} \in H\} = \{ah : \ h \in H\},$$

which we write for simplicity as aH. This is called a *left coset* of H. It is clear that the distinct left cosets are disjoint and each coset has the same number of elements as the number of elements in H. Since each element of G must be exactly in one of the cosets, we see that the *number of elements in G must be a multiple of the number of elements in H*. Note that in the examples (ii) and (iv) of subgroups as given above, this statement is easily verified.

In this context we can also define a *right coset* of H as the set $Ha = \{ha : h \in H\}$. It is clear that a group is partitioned into right cosets too by introducing the equivalence relation $a \sim b$ if $a^{-1}b \in H$. Sometimes right cosets equal the left cosets but not always. In fact, if H is such a subgroup of G such that every left coset is the same as its corresponding left coset, i.e., $aH = Ha$ for all $a \in G$, then we call H a *normal* subgroup of G. Of course, every subgroup of an *abelian* group, where $a \star b = b \star a$ for every $a, b \in G$, is a normal subgroup. But not every group we encountered here is abelian (for example, $S_3, O(3)$). Another equivalent definition of a normal subgroup is: a subgroup H is a normal subgroup of G if for every $h \in H$, $aha^{-1} \in H$ for all $a \in G$.

An interesting thing about normal subgroups is that we can define a group structure on the set of cosets of G: for cosets $A = aH$ and $B = bH$ we may define $AB = \{g_1g_2 : \ g_1 \in A, \ g_2 \in B\}$. It can be shown **(Exercise)** that AB is the coset abH and $AH = A = HA$. If we write $A_1 = a^{-1}H$, then $A_1A = AA_1 = H$. So the cosets constitute a group under this binary operation which we call the quotient group of G modulo H. We write G/H for this *quotient group*. It is obvious that $\text{order}(G/H) = \text{order}(G)/\text{order}(H)$, if G is a finite group.

We have already seen that if we have a finite set A, then the group of permutations of A constitute a group. If we have a set $\{1, 2, 3\}$ consisting of three elements, then we write the permutation which takes 1 to i_1, 2 to i_2 and 3 to i_3 as $\begin{pmatrix} 1 & 2 & 3 \\ i_1 & i_2 & i_3 \end{pmatrix}$. So, $\begin{pmatrix} 1 & 2 & 3 \\ 2 & 1 & 3 \end{pmatrix}$ stands for the permutation which takes 1 to 2, 2 to 1 but keeps 3 as it is. We may define composition of permutation in the usual way:

$$\begin{pmatrix} 1 & 2 & 3 \\ 2 & 3 & 1 \end{pmatrix} \begin{pmatrix} 1 & 2 & 3 \\ 2 & 1 & 3 \end{pmatrix} = \begin{pmatrix} 1 & 2 & 3 \\ 3 & 2 & 1 \end{pmatrix}.$$

We may take the permutation $\begin{pmatrix} 1 & 2 & 3 \\ 1 & 2 & 3 \end{pmatrix}$ as the identity element.

If we look at the multiplication table of a group, we see that multiplication by a fixed element with every element of a group really permutes the elements of the group. In fact, one can show in this way that *every group G is isomorphic to a subgroup of a group of permutations* of the same set G. Proof is really not difficult. The reader can construct a proof.

In view of the above theorem, permutation groups have a position of importance in the study of the structure of groups. We call a permutation which interchanges only two elements of a set and keeps unchanged every other element a *transposition*. It is clear that a transposition is in fact an element of order 2. It should be clear that *every permutation is a compositon of transpositions*. If a permutation is a composition of an even number of transpositions, then we call the permutation an *even permutation*, otherwise an odd permutation. No matter how we interchange elements to achieve a definite permutation, the number of such interchanges has the same parity.

Suppose p is a permutation of a set S. For a given $a \in S$, $pa \in S$, $p^2a \in S$, etc. There is a smallest positive integer l depending on a such that $p^l a = a$. The *orbit* of a under p consists of the set $\{a, pa, p^2a, \cdots, p^{l-1}a\}$. By a *cycle* of p we mean the ordered l-tuple $(a, pa, p^2a, \cdots, p^{l-1}a)$. It is clear that we know the permutation p if we know all its cycles. Look at the permutation

$$p = \begin{pmatrix} 1 & 2 & 3 & 4 & 5 & 6 \\ 2 & 1 & 3 & 5 & 6 & 4 \end{pmatrix}.$$

We see that $p1 = 2, p2 = 1, p3 = 3, p4 = 5, p5 = 6, p6 = 4$. So the cycles of p are (1 2)(3) and (4 5 6). We would like to use the term *cycle* for a permutation $(i_1\ i_2\ \cdots\ i_{r-1}\ i_r)$ which takes i_1 to i_2, i_2 to $i_3, \cdots, i_{r-1}$ to i_r and i_r to i_1 and keeps all other elements fixed. Thus, if on $S = \{1\ 2\ 3\ 4\ 5\ 6\ 7\ 8\ 9\}$ by (1 3 4 2 6) we mean the permutation

$$\begin{pmatrix} 1 & 2 & 3 & 4 & 5 & 6 & 7 & 8 & 9 \\ 3 & 6 & 4 & 2 & 5 & 1 & 7 & 8 & 9 \end{pmatrix}.$$

We multiply two cycles in the obvious way. (5 6 4 1 8)(1 2 3) in the set S as above would be

$$\begin{pmatrix} 1 & 2 & 3 & 4 & 5 & 6 & 7 & 8 & 9 \\ 8 & 2 & 3 & 1 & 6 & 4 & 7 & 5 & 9 \end{pmatrix}\begin{pmatrix} 1 & 2 & 3 & 4 & 5 & 6 & 7 & 8 & 9 \\ 2 & 3 & 1 & 4 & 5 & 6 & 7 & 8 & 9 \end{pmatrix}$$

$$= \begin{pmatrix} 1 & 2 & 3 & 4 & 5 & 6 & 7 & 8 & 9 \\ 2 & 3 & 8 & 1 & 6 & 4 & 7 & 5 & 9 \end{pmatrix}.$$

This can be written in terms of cycles as (1 2 3 8 5 6 4)(7)(9). It can be proven that *every permutation is a product of disjoint cycles.*

Let G be a group of permutations of a set S and let $G_s = \{g \in G : \ gs = s\}$, which is easily seen to be a subgroup of G. We call G_s the *stabilizer* of s. Recall our earlier discussion of symmetries of various figures and that symmetry of an equilateral triangle can be deemed as a permutation of its vertices. In fact, an equilateral triangle has six symmetries, each of which corresponds to a permutation of its vertices. The stabilizers of each of its vertices are the reflection about the medians passing through the vertices.

Given $s \in S$, by an *orbit* of s we mean the set $O_s = \{gs : \ g \in G\}$. The idea of looking at orbits is to look at the number of elements in a group. In fact, one has the following *counting principle*: If G is a group of permutations of a set S, then for $s \in S$, the order of G is equal to the number of elements in the stabilizer of s, multiplied by the number of elements in the orbit of s.

We shall use this counting principle to enumerate all the symmetries of certain regular solids. For example, let us enumerate all the rotational symmetries of a regular tetrahedron. Let A, B, C, D be the vertices of the tetrahedron. One of the symmetries is obtained by rotating the tetrahedron about the line passing through the vertex A and the centroid of the triangle BCD, through an angle of 120° so that the configuration *abcd* will go to *acdb*, which is equivalent to the permutation $\begin{pmatrix} A & B & C & D \\ A & C & D & B \end{pmatrix}$. Similarly the permutation

$$\begin{pmatrix} A & B & C & D \\ A & D & B & C \end{pmatrix}$$

is another rotational symmetry of the tetrahedron. It is easily seen that the orbit of any of the vertices has four elements, whereas the stabilizer of any of the vertices has three elements. So the group of rotational symmetries of a regular tetrahedron has 12 elements. In fact, they are

$$\begin{pmatrix} A & B & C & D \\ A & B & C & D \end{pmatrix}, \begin{pmatrix} A & B & C & D \\ A & C & D & B \end{pmatrix}, \begin{pmatrix} A & B & C & D \\ A & D & B & C \end{pmatrix},$$
$$\begin{pmatrix} A & B & C & D \\ C & B & D & A \end{pmatrix}, \begin{pmatrix} A & B & C & D \\ D & B & A & C \end{pmatrix}, \begin{pmatrix} A & B & C & D \\ B & D & C & A \end{pmatrix},$$
$$\begin{pmatrix} A & B & C & D \\ D & A & C & B \end{pmatrix}, \begin{pmatrix} A & B & C & D \\ B & C & A & D \end{pmatrix}, \begin{pmatrix} A & B & C & D \\ C & A & B & D \end{pmatrix},$$
$$\begin{pmatrix} A & B & C & D \\ B & A & D & C \end{pmatrix}, \begin{pmatrix} A & B & C & D \\ C & D & A & B \end{pmatrix}, \begin{pmatrix} A & B & C & D \\ D & C & B & A \end{pmatrix}.$$

In this context it is interesting to mention the following.

THEOREM: *Every finite subgroup G of $SO(3)$, the group of proper rotations in $\mathbb{R}^3$, is one of the following:*

(i) *C_k: the group of rotations by multiples of $\frac{2\pi}{k}$ about a line for some $k \in \mathbb{N}$,*

(ii) *D_ℓ: the dihedral group of symmetries of a regular polygon of ℓ sides.*

(iii) *T: the tetrahedral group of twelve rotations, carrying a regular tetrahedron to itself.*

(iv) *O: the octahedral group of order 24 of rotations of a cube, or of a regular octahedron.*

(v) *I: the icosahedral group of 60 rotations of a regular dodecahedron or a regular icosahedran* (Figure 7.18).

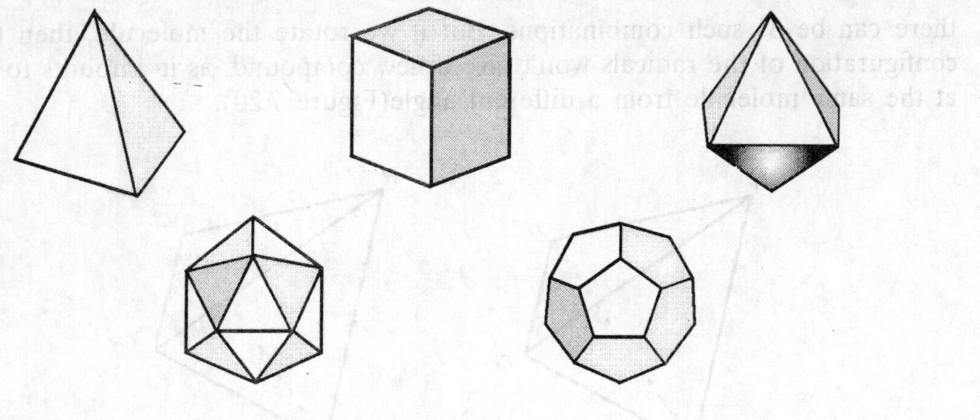

Figure 7.18

The solids tetrahedron, cube, octahedron, dodecahedron and icosahedron are the only regular polyhedra, called the platonic solids, in each case of which the faces are of the same kind of regular polygons. For example, a tetrahedron has faces which are equilateral triangles, whereas a cube has squares. Its proof is by a deep application of group theory, and interested readers may refer [1].

We conclude this section with an application of group theory to combinatorics. Let us start with an illustration (Figure 7.19).

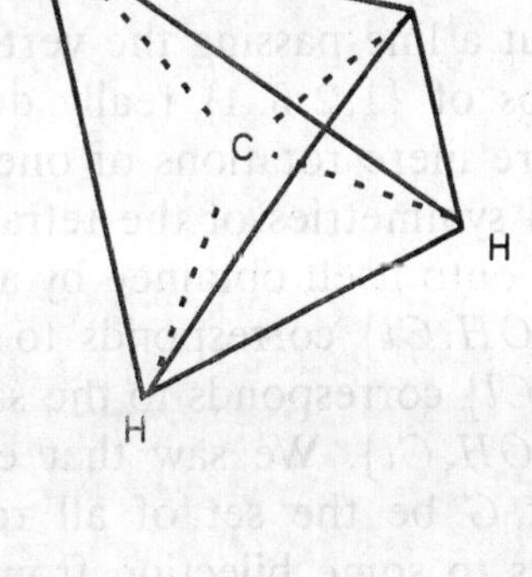

Figure 7.19

It is known in chemistry that methane has a tetrahedral structure with carbon sitting at the centroid of a regular tetrahedron and hydrogen atoms attached to the four valence bonds at the vertices. We know further that the hydrogen atoms can be replaced by monovalent elements or radicals, e.g., Cl, OH, CH_3, etc. Now suppose we have a choice of three monovalent radicals H, Cl, OH (say) to be placed at the vertices of the tetrahedron, how many different choices can we get? Ordinarily we should expect that there would be 81 different compounds. If we number the vertices 1,2,3,4, then each of the vertices has three choices to be occupied. So there can be 3^4 such combinations. But if we rotate the molecule, then the new configuration of the radicals won't give a new compound, as it amounts to looking at the same molecule from a different angle(Figure 7.20).

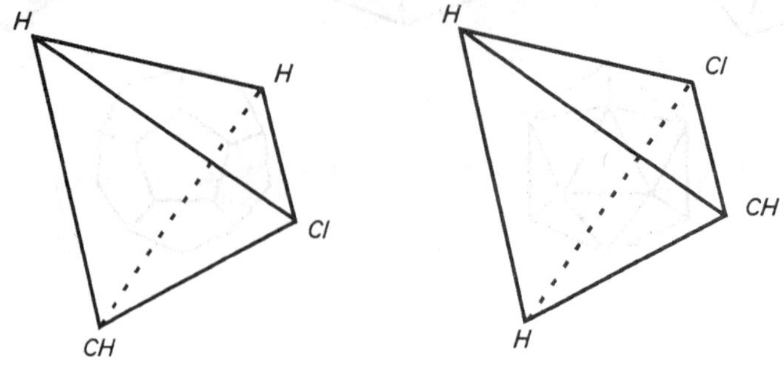

Figure 7.20

In fact, each compound formed corresponds to a map

$$f: \{1,2,3,4\} \longrightarrow \{H, Cl, OH\} .$$

The map f given by $f(1) = H, f(2) = OH, f(3) = Cl, f(4) = H$ corresponds to the compound in which the vertex 1 is occupied by H, 2 by OH, 3 by Cl and 4 by H. It obviously does not differ from the compound where the vertex 1 is occupied by H, 2 by H, 3 by OH, and 4 by Cl as it is obtained by rotating the molecule through an angle of 120° about a line passing the vertex 1 and the centroid of the opposite face. So all the maps of $\{1,2,3,4\}$ really do not give new compounds. Many of the 81 possibilities are mere rotations of one another.

If p is one of the rotational symmetries of the tetrahedron with vertices 1,2,3,4, then it is a bijection $\{1,2,3,4\}$ onto itself obtained by a rotation of the tetrahedron. If map $f : \{1,2,3,4\} \longrightarrow \{H, OH, Cl\}$ corresponds to a compound, then the map $f \circ p : \{1,2,3,4\} \longrightarrow \{H, OH, Cl\}$ corresponds to the same compound. Let us write $V = \{1,2,3,4\}$ and $R = \{H, OH, Cl\}$. We saw that each compound corresponds to a map from V to R. Let G be the set of all rotational symmetries of the tetrahedron which corresponds to some bijection from V to V. Now if f_1 and f_2 are two functions from V to R and there is a $p \in G$ such that $f_1 \circ p = f_2$, then clearly f_1 and f_2 correspond to the same compound. This can be seen in a different light as follows.

Let R^V stand for the set of all maps from $V \longrightarrow R$ and G be the group of all rotational symmetries of V. Let us define a relation $\sim$ in R^V by $f_1 \sim f_2$, if there is an element $p \in G$ such that $f_1 \circ p = f_2$. It is really not hard to show that $\sim$ is an equivalence relation on R^V. In fact, reflexivity is the same as the existence of identity in G, symmetry follows from the existence of inverse in G, while transitivity follows from the fact that group operation is defined in G. The equivalence relation $\sim$ in R^V decomposes it into disjoint equivalence classes. As we had said before, an equivalence class corresponds to a distinct compound and we have to enumerate the disjoint equivalence classes into which R^V is broken up.

It has been observed earlier that the group of symmetries of a tetrahedron is a group of permutations of its vertices. Given a group of permutations G, we define the *cycle index* of the group by a polynomial in many variables as:

$$P(G\ ;\ x_1, x_2, \cdots, x_n) = \frac{1}{|G|}\sum_{p \in G} x_1^{k_1} x_2^{k_2} \cdots x_n^{k_n},$$

where p is a permutation with k_1-cycles of length 1, k_2-cycles of length 2,$\cdots$, k_n-cycles of length n, and $|G|$ stands for the number of elements in G.

To understand this let us look at the group of symmetries of an equilateral triangle with vertices 1,2,3. We saw earlier that the group of symmetries of the triangle coincides with the group of all permutations of $\{1, 2, 3\}$ and the element $\begin{pmatrix} 1 & 2 & 3 \\ 1 & 2 & 3 \end{pmatrix}$ consists of three cycles of length one and no other cycles, whereas $\begin{pmatrix} 1 & 2 & 3 \\ 2 & 3 & 1 \end{pmatrix}$ has one cycle of length 3 $\begin{pmatrix} 1 & 2 & 3 \\ 2 & 1 & 3 \end{pmatrix}$, consists of one cycle of length 1 and one cycle of length 2 and so on. One can easily compute the cycle index of the equilateral triangle to be the polynomial

$$\frac{1}{6}(x_1^3 + x_3 + x_3 + x_1x_2 + x_1x_2 + x_1x_2) = \frac{1}{6}(x_1^3 + 3x_1x_2 + 2x_3).$$

The group of rotational symmetries of a tetrahedron with the set of vertices $\{1, 2, 3, 4\}$ is the group of permutations consisting of

$$\begin{pmatrix} 1 & 2 & 3 & 4 \\ 1 & 2 & 3 & 4 \end{pmatrix}, \begin{pmatrix} 1 & 2 & 3 & 4 \\ 2 & 3 & 1 & 4 \end{pmatrix}, \begin{pmatrix} 1 & 2 & 3 & 4 \\ 3 & 1 & 2 & 4 \end{pmatrix}, \begin{pmatrix} 1 & 2 & 3 & 4 \\ 3 & 2 & 4 & 1 \end{pmatrix},$$
$$\begin{pmatrix} 1 & 2 & 3 & 4 \\ 4 & 2 & 1 & 3 \end{pmatrix}, \begin{pmatrix} 1 & 2 & 3 & 4 \\ 2 & 4 & 3 & 1 \end{pmatrix}, \begin{pmatrix} 1 & 2 & 3 & 4 \\ 4 & 1 & 3 & 2 \end{pmatrix}, \begin{pmatrix} 1 & 2 & 3 & 4 \\ 1 & 3 & 4 & 2 \end{pmatrix},$$
$$\begin{pmatrix} 1 & 2 & 3 & 4 \\ 1 & 4 & 2 & 3 \end{pmatrix}, \begin{pmatrix} 1 & 2 & 3 & 4 \\ 2 & 1 & 4 & 3 \end{pmatrix}, \begin{pmatrix} 1 & 2 & 3 & 4 \\ 3 & 4 & 1 & 2 \end{pmatrix}, \begin{pmatrix} 1 & 2 & 3 & 4 \\ 4 & 3 & 2 & 1 \end{pmatrix},$$

and its cycle index is given by $(1/12)(x_1^4 + x_1x_3 + x_1x_3 + x_1x_3 + x_1x_3 + x_1x_3 + x_1x_3 + x_1x_3 + x_1x_3 + x_2^2 + x_2^2 + x_2^2) = (1/12)(x_1^4 + 8x_1x_2 + 3x_2^2)$.

The reader is invited to calculate the cycle indices of other groups of symmetries of the five platonic solids. With this background, we state the following.

POLYA ENUMERATION THEOREM: *Let D be a finite set and G a group of* *permutations of D. Let R be another finite set with* m *elements. Define the equivalence relation in* R^D *by* $f \sim g$ *if there is a* $p \in G$ *such that* $f \circ p = g$. *Let* $Z(G\ ;\ x_1, x_2, \cdots, x_n)$ *be the cycle index of the group G. The number of disjoint equivalence classes of* R^D *is equal to* $Z(G\ ;\ m, m, m, \cdots, m)$.

Now we go back to our problem of the number of choices of placing the three radicals H, Cl, OH at the vertices of a tetrahedron. By the Polya enumeration theorem, there will be

$$\frac{1}{12}(81 + 8 \cdot 3 \cdot 3 + 3 \cdot 3^2) = \frac{1}{12} \cdot 180 = 15$$

compounds formed by the radicals H, OH, Cl at the ends of the four valence bonds of carbon.

Similarly, let $D = \{x_1, x_2, x_3\}$, $R = \{y_1, y_2, y_3, y_4\}$ and G be the group of all permutations of D. We have already seen that the cycle index of G is $(1/6)(x_1^3 + 3x_1x_2 + 2x_3)$. So the number of equivalence classes of R^D is $(1/6)(64+48+8) = 20$. In fact, this number 20 explains why there are only 20 naturally occuring amino acids, since amino acids are formed by combining any three of the four constituent nucleotides with repetitions allowed, with a permutation of the nucleotides giving the same amino acid.

7.2 Rings and Fields

We have seen several examples of sets with two binary operations, addition and multiplication. These are integers, rational numbers, real numbers and the set of all linear transformations of $\mathbb{R}^2$. (In the last example, multiplication is to be understood as composition). Each of these is an example of a *ring*, which we formally define below.

A non-empty set R together with two binary operations called addition (+) and multiplication (.) is called a ring if

(i) $a + (b + c) = (a + b) + c$, for all $a, b, c \in R$;

(ii) there exists a unique element $0 \in R$ such that $a + 0 = 0 + a = a$;

(iii) for all $a \in R$, there exists $b \in R$ such that $a + b = b + a = 0$;

(iv) $a + b = b + a$ for all $a, b \in R$;

(v) $a \cdot (b \cdot c) = (a \cdot b) \cdot c$ for all $a, b, c \in R$;

(vi) $a \cdot (b + c) = a \cdot b + a \cdot c$ for all $a, b, c \in R$

(vii) $(a+b)\cdot c = a\cdot c + b\cdot c$ for all $a, b, c \in R$.

The element 0 is called the additive identity. A ring $(R, +, \cdot)$ is said to be commutative if

(viii) $a \cdot b = b \cdot a$ for all $a, b \in R$.

If there exists an element $c \in R$ such that

(ix) $a\cdot c = c\cdot a = a$ for all $a \in R$, then c is unique and is called the multiplicative identity (denoted by 1) and in such a case $(R, +, \cdot)$ is said to be a ring with identity.

The integers, rational numbers and real numbers are commutative rings with identity. The set $\mathcal{L}(\mathbb{R}^2)$ of linear transformations on $\mathbb{R}^2$ is a ring with identity. This is not a commutative ring.

A commutative ring with identity $(R, +, \cdot)$ is said to be a *field* if

(x) for all $a \in R$, $a \neq 0$, there exists $b \in R$ such that $a \cdot b = b \cdot a = 1$.

As noted in Chapter 2, rational numbers and real numbers are fields. Integers of course do not form a field. Note that when $(R, +, \cdot)$ is a ring, $(R, +)$ is an abelian group **(Exercise)**.

We will now give examples of finite rings. For an integer $n > 1$, let

$$\mathbb{Z}_n = \{0, 1, 2, \cdots, n-1\}.$$

Define binary operations $+_n$ and $\cdot_n$ (read as addition mod n and multiplication mod n respectively) as follows: For $a, b \in \mathbb{Z}_n$,

$$a +_n b = r,$$

where r $(0 \leq r \leq n-1)$ is the remainder when $a + b$ is divided by n. Likewise,

$$a \cdot_n b = s,$$

where $s (0 \leq s \leq n-1)$ is the remainder when $a\cdot b$ is divided by n. Here, unlike in the case of integers, one can have a non-zero element $a \in \mathbb{Z}_n$ such that $a^2 \equiv a \cdot_n a = 0$. As an example consider $a = 3$ and $n = 9$. Henceforth, we shall replace $+_n$ and $\cdot_n$ by $+$ and $\cdot$ respectively when there is no confusion.

It is easy to verify that $(\mathbb{Z}_n, +, \cdot)$ is a commutative ring with identity **(Exercise)**. Also, $k \in \mathbb{Z}_n$ has a multiplicative inverse if and only if the highest common factor (h.c.f) of k and n is 1. This can be deduced using *Euclid's algorithm*: the h.c.f. of k and n is 1 if and only if there exist integers x, y such that

$$kx + ny = 1.$$

It then follows that if p is a prime number, then $(\mathbb{Z}_p, +, \cdot)$ is a field.

Exercises

(i) In the field $\mathbb{Z}_7$, compute the multiplicative inverses of 2, 3 and 4.

(ii) Let $U_n = \{k \in \mathbb{Z}_n : \text{there exists } j \in \mathbb{Z}_n \text{ such that } k \cdot j = 1\}$. Show that $(U_n, \cdot)$ is a group.

It can be shown that if F is a finite field (i.e., if F is a finite set and for some binary operations $+, \cdot$ on F, $(F, +, \cdot)$ is a field), then the number of elements of F (called the order of the field F) is a power of a prime. The particular infinite field $\mathbb{Q}$ and the finite fields $\mathbb{Z}_p$ (for primes p) are of special interest and are called *prime fields*.

Examples

(i) Rings of polynomial functions: A real polynomial P is a function from $\mathbb{R}$ to $\mathbb{R}$ such that

$$P(x) = \sum_{k=0}^{n} a_k x^k \text{ for some integer } n \geq 0 \text{ and } a_0, a_k, \cdots a_n \in \mathbb{R}.$$

The integer n is called the *degree* of the polynomial P if $a_n \neq 0$.

If P_1 and P_2 are two polynomials of degree n_1 and n_2 respectively, we define their sum and product (by pointwise sum and product) as

$$(P_1 + P_2)(x) = P_1(x) + P_2(x) \text{ and } (P_1 P_2)(x) = P_1(x) P_2(x).$$

Then it is easy to see that $P_1 + P_2$ and $P_1 P_2$ are again polynomials. While the degree of the first is less than or equal to $\max\{n_1, n_2\}$, that of the second is $n_1 + n_2$.

If we denote by R the set of all polynomials, then one can easily verify that $(R, +, \cdot)$ is a commutative ring with identity.

(ii) Rings of continuous functions: Let S be the set of all continuous functions from $\mathbb{R}$ to $\mathbb{R}$. The sum and product in S is defined pointwise (as in the case of polynomials). Then $(S, +, \cdot)$ is a commutative ring with identity.

(iii) Matrix rings: Fix an integer $n \geq 1$. Let

$$M_n = \{A : A \text{ is an } n \times n \text{ matrix with integer or real entries } \}.$$

For $A = ((a_{ij}))$, $B = ((b_{ij})) \in M_n$, $1 \le i \le n$, $1 \le j \le n$, let $C = ((c_{ij}))$ and $D = ((d_{ij}))$ be defined as

$$c_{ij} = a_{ij} + b_{ij} \quad \text{and} \quad d_{ij} = \sum_{k=1}^{n} a_{ik} b_{kj}, \quad 1 \le i, j \le n.$$

The matrix C is defined to be the sum of A and B ($C = A+B$) and D is defined to be the product of A and B ($D = AB$). We call the matrices I and O, the identity and zero matrices respectively, if their elements are given respectively as

$(I)_{ij} = 1$ if $i = j$ and $= 0$ otherwise, and $(O)_{ij} = 0$ for all i, j.

As in the case $n = 2$, one can identify matrices in M_n with the linear transformation on $\mathbb{R}^n$. The product of two matrices then corresponds to composition of the corresponding transformations (see Chapter 5).

It can be verified that $(M_n, +, \cdot)$ is a ring with identity. This ring is non-commutative (since for two $n \times n$ matrices A and B, AB is not necessarily equal to BA.) (**Exercise**: For $n = 2$, find examples of A and B with such a property.)

As in the case of the ring $\mathbb{Z}_n$, here also one can have a matrix A such that $A^2 = 0$. For example, for $n = 2$ take $A = \left(\begin{smallmatrix} 0 & 1 \\ 0 & 0 \end{smallmatrix}\right)$, in which case it is easy to see that $A^2 = \left(\begin{smallmatrix} 0 & 0 \\ 0 & 0 \end{smallmatrix}\right)$.

(iv) Boolean rings: Let X be a non-empty set and let $R = \mathcal{P}(X)$—the set of all subsets of X (also called the *power set* of X). We define $+$ and $\cdot$ on R as follows: for $A, B \in R$, let

$$A + B = (A \cap B^c) \cup (A^c \cap B)$$

and

$$A \cdot B = A \cap B.$$

The sum $A + B$ defined above is known as the *symmetric difference* of A and B and is also denoted by $A \Delta B$. Then $(R, +, \cdot)$ is a commutative ring with identity. Here the additive identity 0 is the empty set $\emptyset$ and the multiplicative identity $1 = X$.

This ring has a property that distinguishes it from many of the examples considered earlier:

$$A^2 = A \quad \text{for all} \quad A \in R.$$

Any ring R in which $x^2 = x$ for all $x \in R$ is called a *Boolean ring* and so $R = \mathcal{P}(X)$ is Boolean.

Exercise: Show that any Boolean ring R is commutative and that $2x = x + x = 0$ for all $x \in R$.

Note that the power set $\mathcal{P}(X)$ of a set along with the operations of unions and intersections does not form a ring (why?).

(v) Complex numbers: $\mathbb{C} = \{a + ib | a, b \in \mathbb{R}\}$ is a field.

(vi) Gaussian numbers: $\mathbb{Q}(i) = \{r + is | r, s \in \mathbb{Q}\}$ can be easily verified to be a field and is a subfield of $\mathbb{C}$.

(vii) Rational quadratic surds: $\mathbb{Q}(\sqrt{2}) = \{r + s\sqrt{2} | r, s \in \mathbb{Q}\}$ is a subfield of $\mathbb{R}$.

(viii) Formal polynomials over a ring: Let R be a ring and let X be an indeterminate (or just a dummy symbol). We denote by $R[X]$ the set of all formal polynomial expressions in X with coefficients in R:

$$R[X] = \{a_0 + a_1 X + \cdots + a_n X^n | a_0, a_1, \cdots, a_n \in R, \ n \in \mathbb{N} \cup \{0\}\}.$$

The integer n is called the degree of the polynomial $a_0 + a_1 X + \cdots + a_n X^n$ if $a_n \neq 0$. Two polynomials in $R[X]$ are said to be equal if their degrees are equal and their coefficients are equal. It is easy to verify that the sum and product of two polynomials (defined exactly as in example (i)) are again polynomials. If R has 1, a non-zero polynomial whose coefficient a_n of the term of the highest degree X^n is 1, it is called a *monic* polynomial. Polynomials of degree $0, 1, 2, 3, 4, 5, \cdots$ are called *constant, linear, quadratic, cubic, biquadratic, quintic,* $\cdots$ respectively.

(ix) Formal polynomials in two variables: Given a ring R, one defines a polynomial

$$p(X, Y) = \sum_{k=0}^{n} \sum_{j=0}^{m} a_{j.k} X^j Y^k$$

for $a_{j,k}$ in R. Check (**Exercise**) that

$$R[X, Y] = \left\{ \begin{array}{c} p(X, Y) : p(X, Y) \text{ is a polynomial in two indeterminates} \\ \text{with coefficients in } R \end{array} \right\}$$

is a ring. For the polynomial

$$p(X,Y) = \sum_{k=0}^{n} \sum_{j=0}^{m} a_{j,k} X^j Y^k,$$

ℓ is called the degree of the polynomial $p(X,Y)$ if $\ell = \max\{j+k : a_{jk} \neq 0\}$. A non-zero polynomial

$$\sum_{\substack{j+k=\ell \\ j,k\geq 0}} a_{j,k} X^j Y^k$$

is called a homogeneous polynomial of degree ℓ.

Application to coding theory

Consider the following scheme for coding a secret message. Let $\mathcal{C}$ consist of the letters $A, B, C, \cdots, Z$ along with $*$, where $*$ represents a space (i.e., gap between two words). Let

$$f : \mathcal{C} \to \mathbb{Z}_{27}$$

be defined by $f(\star) = 0,\ f(A) = 1, f(B) = 2, \cdots, f(Y) = 25,\ f(Z) = 26.$

Thus given a message, applying f to each character, we get a string of numbers, say $a_1, a_2, \cdots, a_m$, in $\mathbb{Z}_{27}$.

Choose elements $p, q \in \mathbb{Z}_{27}$ such that p admits a multiplicative inverse, r (i.e., $p \cdot r = 1$ in $\mathbb{Z}_{27}$). Let

$$b_i = p \cdot (a_i + q), \qquad 1 \leq i \leq m.$$

Let us transmit the message

$$f^{-1}(b_1) f^{-1}(b_2) \cdots f^{-1}(b_m).$$

The person receiving the message should know p, q (which can be agreed upon earlier or conveyed independently). Then he can first compute b_i and then

$$a_i = rb_i - q$$

can be computed.

The original message is then

$$f^{-1}(a_1) \cdots f^{-1}(a_m).$$

Note that in order to "decode", one needs to know the method of coding and also the "seeds" p, q.

Consider the message

ATTACK AT DAWN.

This we write as a string of characters

ATTACK * AT * DAWN

Next the string is transformed as elements of $\mathbb{Z}_{27}$ as

$$1, 20, 20, 1, 3, 11, 0, 1, 20, 0, 4, 1, 23, 14.$$

Let us choose $p = 4,\ q = 16$. Then $b_i = 4(a_i + 16)$ $(1 \leq i \leq 14)$ is the sequence

$$14, 9, 9, 14, 22, 0, 10, 14, 9, 10, 26, 14, 21, 12.$$

The corresponding character sequence (the message that will be transmitted) is

NIINV * JNIJZNUL.

This will be meaningless and only the persons knowing the method of coding and the values of p and q, namely $p = 4,\ q = 16$ would be able to decode; first get b_i from the message received, then get

$$a_i = 7 \cdot b_i - 16$$

and finally, the original message as

$$f(a_1), \cdots, f(a_{14}).$$

It should be mentioned that with the availability of modern computers it is easy to break the code even if one does not know p, q: simply for each choice of p, q (at most 27×27 choices) the computer can decode and then discard outputs that do not make sense. The above example is a rather simplistic exposition of coding. In practice, more sophisticated coding/decoding algorithms are used.

7.3 Algebraic numbers and Algebraic geometry

Let $R[X]$ be the ring of polynomials over a ring R as in Example (viii) in Section

7.2. For $f(X) = a_0 + a_1X + \cdots + a_nX^n \in R[X]$, an element $\alpha \in R$ is called a *root or zero* of $f(X)$ in R if

$$f(\alpha) = a_0 + a_1\alpha + a_2\alpha^2 + \cdots + a_n\alpha^n = 0.$$

For example, take $R = \mathbb{Z}_{16}$ and $f(X) = 8X$, which is a linear expression. How many distinct roots does it have? It is not difficult to verify (**Exercise**) that $0, 2, 4, 6, 8, 10, 12$ and 14 are the roots of f in R. Similarly, $f(X) = X^2 + 1$ over $R = \mathbb{C}$ has exactly two roots i and $-i$.

A complex number α is called an *algebraic number* if it is a root of a non-constant polynomial $p(X)$ with integer coefficients, i.e., $p(\alpha) = 0$, where $p(X) \in \mathbb{Z}[X]$ with degree of p non-zero. We note that a root of a polynomial with rational coefficients is in fact a root of one with integer coefficients. It is known that the set of all algebraic numbers form a subfield of $\mathbb{C}$ and is denoted as $\overline{\mathbb{Q}}$.

There is a deep theorem due to Gauss (which uses many important results from the theory of functions of a complex variable) which says that every non-constant polynomial in $\mathbb{C}[X]$ has a root in $\mathbb{C}$. This is called the **Fundamental Theorem of Algebra.**

The complex numbers which are not algebraic are called *transcendental numbers.* For example, $\sqrt{2}, 3\sqrt{5/7}, \pm i4\sqrt{2}$ are algebraic numbers while $e, \pi, 2^{\sqrt{2}}$ are transcendental. A delightful introduction to these and some other questions discussed in this section can be found in [2].

As a special subset of algebraic numbers, we now consider what are called *constructible numbers*. By this we mean those numbers which can be obtained as lengths of line segments, using the straight edge and compass only. We have already seen in Chapter 2 how the integers and the rationals are constructed in this manner. It is not difficult to see that if a and b are constructible numbers, then so are $a+b$, $a-b$, ab and a/b. For this it is enough to show that ab is constructible and this can be achieved by using the properties of similar triangles as in the left-hand side figure in Figure 7.21.

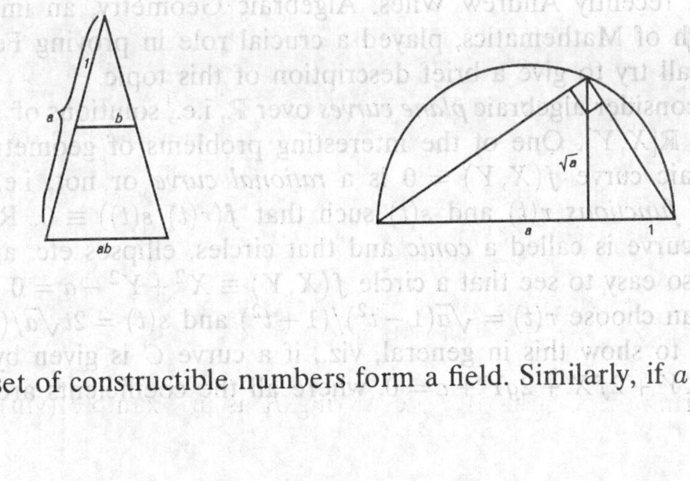

Figure 7.21

Thus the set of constructible numbers form a field. Similarly, if a is constructible, then $\sqrt{a}$ is one such as it follows from the similarity of trangles in the right-hand side figure in Figure 7.21 (**Exercise**).

The numbers like $\sqrt{2}, \sqrt{5+\sqrt{3}}$ are constructible. In fact, all the constructible numbers are obtained as the points of intersection of a circle, given by $(x-\alpha)^2 + (y-\beta)^2 = r^2$, and a straight line, given by $y = mx+c$, with α, β, r, m, c constructible. Gauss showed that a regular 17-gon can be constructed this way by obtaining an expression for $\cos(2\pi/17)$ as a succession of three square roots. Also some of the classical problems of Euclidean geometry, viz, the doubling of cube and trisection of an angle can be shown to involve numbers which are not constructible (see [3]).

A field F is said to be *algebraically closed* if every non-constant polynomial in $F[X]$ has a root in F. Thus by the fundamental theorem of algebra $\mathbb{C}$ is algebraically closed. This is equivalent to saying that the only *irreducible* polynomials in $\mathbb{C}[X]$ are linear polynomials, i.e., every non-constant polynomial in $\mathbb{C}[X]$ is a product of linear ones. This is not true in $\mathbb{R}$ and $\mathbb{Q}$ since, for example, X^2+1 cannot be factored into linear ones with either real or rational coefficients. On the other hand, it is known that $\bar{\mathbb{Q}}$ is algebraically closed.

Algebraic geometry is the study of the zeros of polynomial equations in finitely many variables over a field F. In the case of one variable, this leads to the field of algebraic numbers as we have seen already. If the number of variables is two and the field is $\mathbb{R}$, then we have the familiar plane cartesian geometry, while for $F = Q$, we have the theory of *Diophantine* equations.

Let us begin with Fermat's polynomial $f(X,Y,Z) = X^n + Y^n - Z^n$, homogeneous of degree n in 3 variables. What are the solutions of $f(X,Y,Z) = 0$ in $\mathbb{R}^3$? If $n = 1$, any solution is of the form $(a, b, a+b)$, where a and b are arbitrary real numbers. If $n = 2$, solutions are the so called Pythagorean triples, namely, $(a^2-b^2, 2ab, a^2+b^2)$ for a and b arbitrary real numbers. Notice that in these two cases, if we choose a and b integers, we get the solutions to be all integer triples. What about the case when $n \geq 3$? The famous Last Theorem of Fermat stated (without proof) that there are no integer solutions (a, b, c) with $abc \neq 0$ if $n \geq 3$. This statement remained unproved for over 350 years despite efforts by many over the years and has been proved recently Andrew Wiles. Algebraic Geometry, an important and interesting branch of Mathematics, played a crucial role in proving Fermat's Last Theorem. We shall try to give a brief description of this topic.

Now we shall consider algebraic *plane curves* over $\mathbb{R}$, i.e., solutions of $f(X,Y) = 0$, where $f(X,Y) \in \mathbb{R}[X,Y]$. One of the interesting problems of geometry is to find out if an algebraic curve $f(X,Y) = 0$ is a *rational curve* or not, i.e., if we can find two *rational functions* $r(t)$ and $s(t)$ such that $f(r(t), s(t)) \equiv 0$. Recall that a quadratic plane curve is called a *conic* and that circles, ellipses etc. are examples of conics. It is also easy to see that a circle $f(X,Y) \equiv X^2+Y^2-a = 0$ is a rational curve since we can choose $r(t) = \sqrt{a}(1-t^2)/(1+t^2)$ and $s(t) = 2t\sqrt{a}/(1+t^2)$. It is not very difficult to show this in general, viz., if a curve C is given by $f(X,Y) \equiv aX^2 + bY^2 + 2hXY + 2fX + 2gY + c = 0$, where all the coefficients are real and if the curve C does not degenerate into a pair of straight lines, then C is a rational curve. To see this, let (α, β) be a point on C, set $Y = \beta + t(X - \alpha)$ and eliminating Y we get a quadratic in X: $R(t)X^2 + S(t)X + T(t) = 0$, where R, S and T are polynomials in t. Since $X = \alpha$ is a root of this equation, the other root is given by $X = -\alpha - (S(t)/R(t))$. By substituting one can find a similar rational expression for Y. Geometrically, this means that we are looking at the two points of intersection between the curve C and a line through (α, β) with slope t.

In this argument we have tacitly assumed that a line and a conic C intersect at two points, which is not true if C degenerates into a pair of straight lines. This fact is a consequence of the next more general theorem.

THEOREM (BEZOUT): *Suppose $f(X,Y)$ and $g(X,Y)$ are two polynomials of degrees m and n respectively without any common factors of positive degree. Then the algebraic curves $f(X,Y) = 0$ and $g(X,Y) = 0$ intersect at no more than mn points.*

Thus, if $f(X,Y)$ is an irreducible polynomial of degree $n > 1$, then a line can intersect it at no more than n points, and if we fix a point P on the curve through which the line must pass, then the number of points of intersection other than P is $n - d$, where $d \geq 1$.

The point P is said to be *simple* if $d = 1$, and of *multiplicity* d otherwise. For example, when $d = 2$, i.e., when P is a double point, the curve has two branches at P as in the figure below (Figure 7.22).

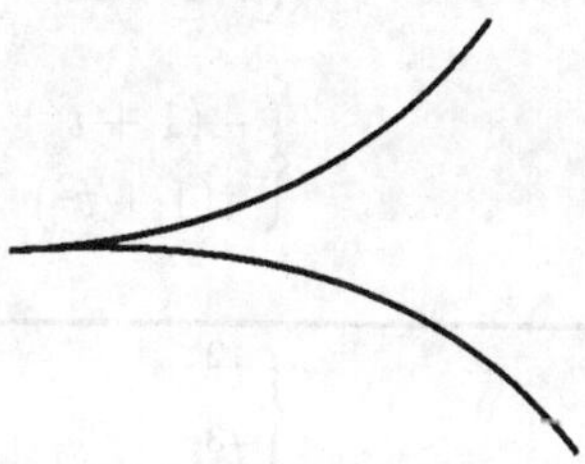

cusp

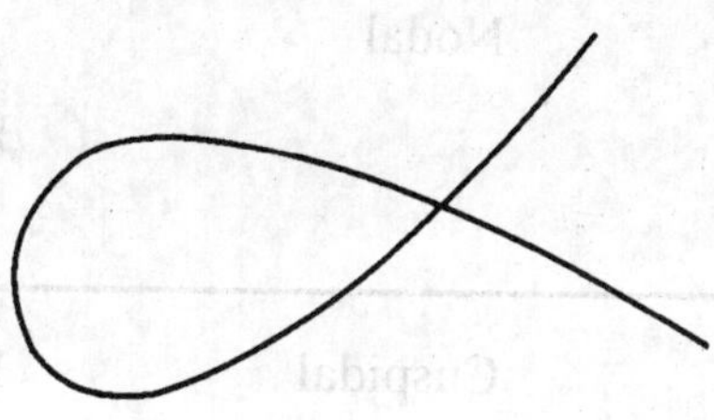

node

Figure 7.22

There can be at most $(n-1)(n-2)/2$ double points on a curve given by an irreducible polynomial of degree n, and if actually δ is the number of such double points on the same curve, then $g \equiv \frac{1}{2}(n-1)(n-2) - \delta$ is called the *genus* of the curve. It is a fact that such a curve is rational if and only if its genus g is zero.

We shall now list some of the well known rational curves.

No.	Curve	$f(X,Y)$	$\begin{cases} r(t) \\ s(t) \end{cases}$
1.	Line	$aX + bY + c,\ a \neq 0$	$\begin{cases} -\dfrac{bt+c}{a} \\ t \end{cases}$
	Parabola	$Y^2 - 4aX$	$\begin{cases} at \\ 2at \end{cases}$
2.	Circle/Ellipse	$(aX)^2 + (bY)^2 - 1$	$\begin{cases} \dfrac{1-t^2}{a(1+t^2)} \\ \dfrac{2t}{b(1+t^2)} \end{cases}$
	Hyperbola	$(aX)^2 - (bY)^2 - 1$	$\begin{cases} \dfrac{1+t^2}{a(1-t^2)} \\ \dfrac{2t}{b(1-t^2)} \end{cases}$
3.	Nodal	$Y^2 - X^2 + X^3$	$\begin{cases} 1+t^2 \\ t(1+t^2) \end{cases}$
		$Y^2 + X^2 + X^3$	$\begin{cases} -(1+t^2) \\ t(1+t^2) \end{cases}$
4.	Cuspidal	$Y^2 - X^3$	$\begin{cases} t^2 \\ t^3 \end{cases}$

It turns out that the notion of a rational curve is very useful in finding solutions of some "Diophantine" equations. For example, knowing that a circle is a rational curve, we can find all the rationals r and s satisfying $r^2 + s^2 = 1$, which is easily seen to be equivalent to finding natural numbers x, y, z satisfying $x^2 + y^2 = z^2$.

In the early seventeenth century, the French mathematician Fermat wondered whether such solutions exist for the equation $x^n + y^n = z^n$ with $n \geq 3$. As before, this can be reduced to looking for rational solutions for the equation $f(X,Y) \equiv X^n + Y^n - 1 = 0$. We can now prove the following.

THEOREM: *Fermat's curve $X^n + Y^n - 1 = 0$ is rational if and only if $n \leq 2$.*

PROOF: We have already seen that for $n \leq 2$, the curve is rational. Now assume that the curve is rational so that it has parametrization $X = p(t)/r(t), Y = q(t)/r(t)$, where p, q and $r(\neq 0)$ are polynomials over $\mathbb{R}$ with degrees a, b, c respectively. We can also assume without loss of generality that (p, r) and (q, r) do not have any common factors. Thus we have

$$p(t)^n + q(t)^n = r(t)^n.$$

Formally differentiating and dividing by n, we get

$$p(t)^{n-1}p'(t) + q(t)^{n-1}q'(t) = r(t)^{n-1}r'(t).$$

Eliminating $q(t)^n$ yields the equality

$$p(t)^{n-1}(q'(t)p(t) - p'(t)q(t)) = r(t)^{n-1}(q'(t)r(t) - q(t)r'(t)).$$

Since p and r have no common factors, $p(t)^{n-1}$ and $r(t)^{n-1}$ must be factors of $(q'(t)r(t) - q(t)r'(t))$ and $(q'(t)p(t) - p'(t)q(t))$ respectively. A comparison of the degrees leads to $a(n-1) \leq (b+c-1)$, and similarly elimination of $p(t)^{n-1}$ and $r(t)^{n-1}$ yields $b(n-1) \leq c+a-1$ and $c(n-1) \leq a+b-1$. All this implies that $(n-1)(a+b+c) \leq 2(a+b+c) - 3$ or $(n-3)(a+b+c) \leq -3 < 0$ and since $a+b+c > 0$, we have $n < 3$. ■

However, we must caution the reader that the above demonstration does not settle the question of Fermat one way or the other because one can possibly have rational points on a curve without the curve being rational.

We shall end this section by giving a list of examples of plane curves of degree higher than 2, with their equations and graphs (Figures. 7.23–7.30).

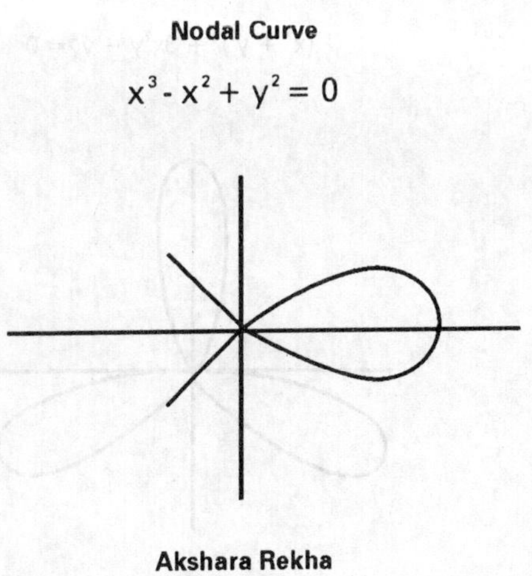

Figure 7.23

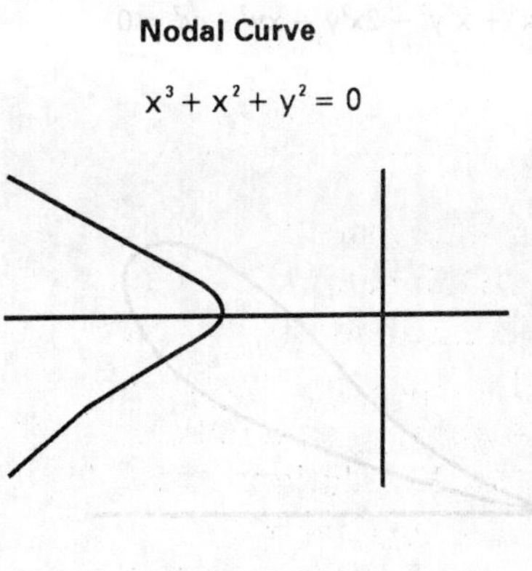

Figure 7.24

Cuspidal Curve

$x^3 - y^2 = 0$

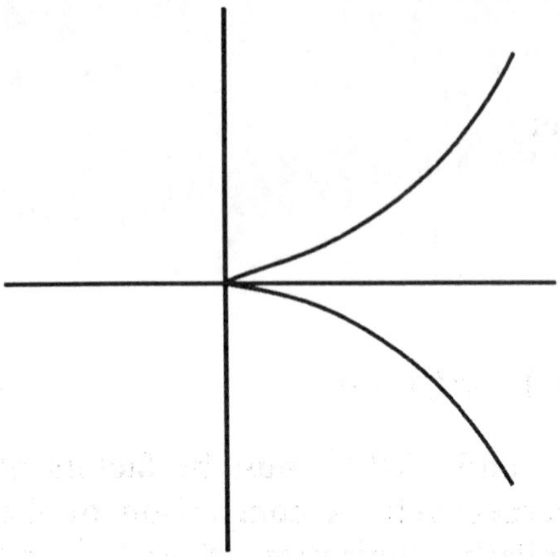

Oonche Neeche Rekha

Figure 7.25

Tacnodal Curve

$2x^4 - 3x^2y + y^2 - 2y^3 + y^4 = 0$

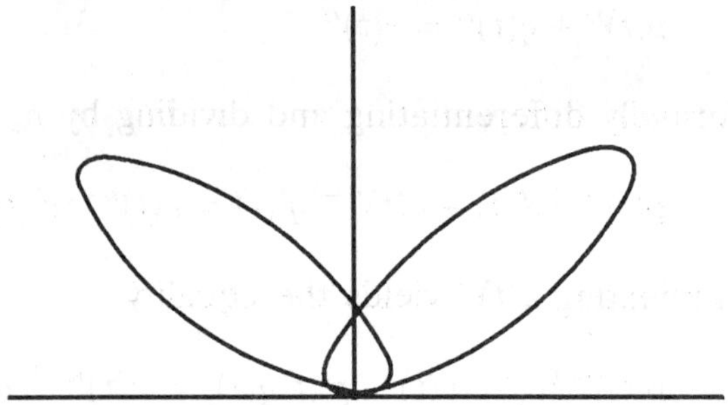

Kharghosh Kan Rekha

Figure 7.26

Ramphold Cuspidal Curve

$x^4 + x^2y^2 - 2x^2y - xy^2 + y^2 = 0$

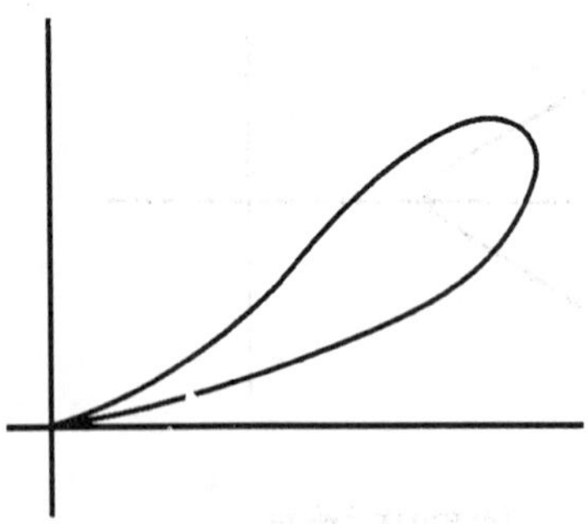

Figure 7.27

Flame Curve

$(x^4 + y^2)^2 + 3x^2y - y^3 = 0$

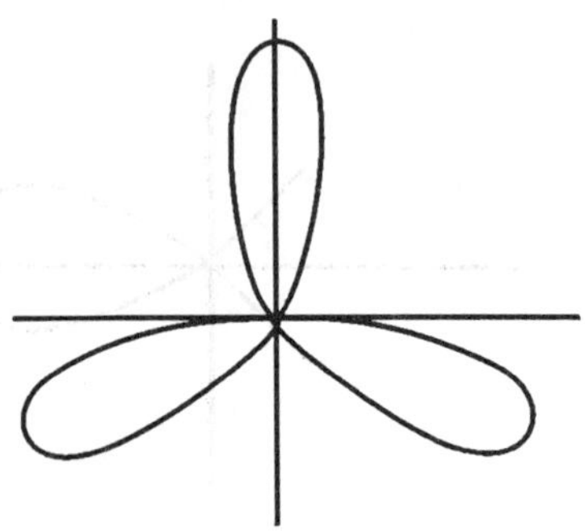

Figure 7.28

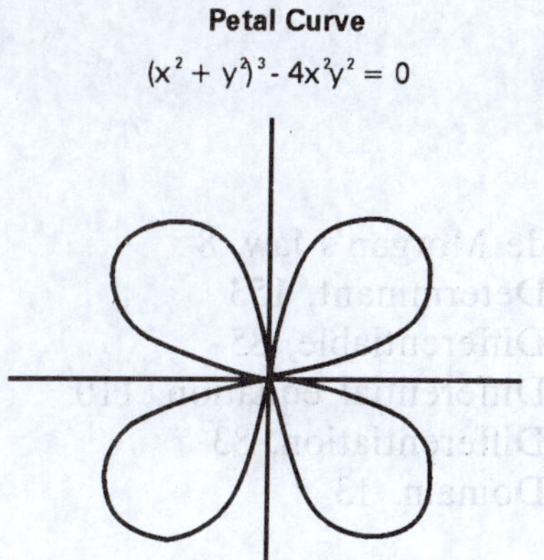

Figure 7.29

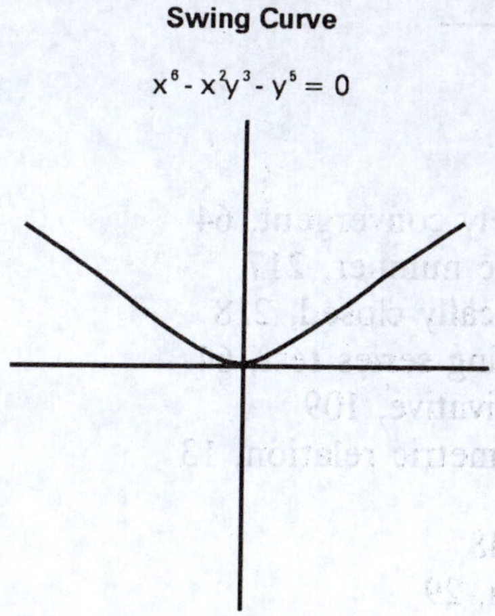

Figure 7.30

Suggested Reading

1. *Introduction to Geometry* by H.S.M. Coxeter, John Wiley, New York.

2. *Algebra in Ancient and Modern Times* by V.S. Varadarajan, Trim series # 14, Hindustan Book Agency, New Delhi 1997.

3. *Famous Problems and other monographs* by F. Klein, W.F. Sheppard, P.A. Macmohan, L.J. Mordell, Chelsea Publishing Co., New York 1980.

4. *A Concrete Introduction to Higher Algebra* by L. Childs, Undergraduate Texts in Mathematics, Springer-Verlag, Heidelberg 1979.

Index